VOLUME ONE HUNDRED AND NINETY FIVE

PROGRESS IN
MOLECULAR BIOLOGY
AND TRANSLATIONAL
SCIENCE

G Protein-Coupled Receptors - Part B

VOLUME ONE HUNDRED AND NINETY FIVE

PROGRESS IN
MOLECULAR BIOLOGY AND TRANSLATIONAL SCIENCE

G Protein-Coupled Receptors - Part B

Edited by

ARUN K. SHUKLA

Department of Biological Sciences and Bioengineering
Indian Institute of technology, Kanpur, India

ACADEMIC PRESS

An imprint of Elsevier

ELSEVIER

Academic Press is an imprint of Elsevier
50 Hampshire Street, 5th Floor, Cambridge, MA 02139, United States
525 B Street, Suite 1650, San Diego, CA 92101, United States
The Boulevard, Langford Lane, Kidlington, Oxford OX5 1GB, United Kingdom
125 London Wall, London EC2Y 5AS, United Kingdom

First edition 2023

Notices
Knowledge and best practice in this field are constantly changing. As new research and experience
broaden our understanding, changes in research methods, professional practices, or medical
treatment may become necessary.

Practitioners and researchers must always rely on their own experience and knowledge in evaluating
and using any information, methods, compounds, or experiments described herein. In using such
information or methods they should be mindful of their own safety and the safety of others, including
parties for whom they have a professional responsibility.

To the fullest extent of the law, neither the Publisher nor the authors, contributors, or editors, assume
any liability for any injury and/or damage to persons or property as a matter of products liability,
negligence or otherwise, or from any use or operation of any methods, products, instructions, or ideas
contained in the material herein.

ISBN: 978-0-323-99434-7
ISSN: 1877-1173

For information on all Academic Press publications
visit our website at https://www.elsevier.com/books-and-journals

Publisher: Zoe Kruze
Acquisitions Editor: Leticia Lima
Developmental Editor: Jhon Michael Peñano
Production Project Manager: Sudharshini Renganathan
Cover Designer: Matthew Limbert

Typeset by STRAIVE, India
Transferred to Digital Printing 2023

Working together
to grow libraries in
developing countries

www.elsevier.com • www.bookaid.org

Contents

Contributors

Ogün Adebali
Molecular Biology, Genetics and Bioengineering Program, Faculty of Engineering and Natural Sciences, Sabanci University, Istanbul; TÜBİTAK Research Institute for Fundamental Sciences, Gebze, Turkey

Sandra Arroyo-Urea
Institute for Biocomputation and Physics of Complex Systems (BIFI) and Laboratorio de Microscopías Avanzadas (LMA), University of Zaragoza, Zaragoza, Spain

Sarah L. Baccetto
College of Pharmacy and Nutrition, University of Saskatchewan, Saskatoon, SK, Canada

Tallan Black
College of Pharmacy and Nutrition, University of Saskatchewan, Saskatoon, SK, Canada

Ángela Carrión-Antolí
Institute for Biocomputation and Physics of Complex Systems (BIFI) and Laboratorio de Microscopías Avanzadas (LMA), University of Zaragoza, Zaragoza, Spain

Amy Davies
Section of Cell Biology and Functional Genomics, Department of Metabolism, Digestion and Reproduction, Imperial College London, London, United Kingdom

Luca Franchini
Department of Pharmacology and Physiology, University of Rochester Medical Center, Rochester, NY, United States

Yang Gao
Department of Cardiology of Sir Run Run Shaw Hospital, Zhejiang University School of Medicine; Liangzhu Laboratory, Zhejiang University Medical Center; Key Laboratory of Cardiovascular Intervention and Regenerative Medicine of Zhejiang Province, Hangzhou, China

Javier García-Nafría
Institute for Biocomputation and Physics of Complex Systems (BIFI) and Laboratorio de Microscopías Avanzadas (LMA), University of Zaragoza, Zaragoza, Spain

Caroline M. Gorvin
Centre of Membrane Proteins and Receptors (COMPARE), Universities of Birmingham and Nottingham; Institute of Metabolism and Systems Research (IMSR) and Centre for Endocrinology, Diabetes and Metabolism, Birmingham Health Partners, University of Birmingham, Birmingham, United Kingdom

Hye Ji J. Kim
College of Pharmacy and Nutrition, University of Saskatchewan, Saskatoon, SK, Canada

Robert B. Laprairie
College of Pharmacy and Nutrition, University of Saskatchewan, Saskatoon, SK;
Department of Pharmacology, College of Medicine, Dalhousie University, Halifax, NS,
Canada

Yanjun Li
Department of Cardiology of Sir Run Run Shaw Hospital, Zhejiang University School of
Medicine; Liangzhu Laboratory, Zhejiang University Medical Center, Hangzhou, China

Ines Liebscher
Rudolf Schönheimer Institute of Biochemistry, Medical Faculty, Leipzig University,
Leipzig, Germany

Hongnan Liu
Department of Cardiology of Sir Run Run Shaw Hospital, Zhejiang University School of
Medicine; Liangzhu Laboratory, Zhejiang University Medical Center, Hangzhou, China

Susruta Majumdar
Center for Clinical Pharmacology, University of Health Sciences & Pharmacy at St Louis and
Washington University School of Medicine, St Louis, MO, United States

Jorge Mallor-Franco
Institute for Biocomputation and Physics of Complex Systems (BIFI) and Laboratorio de
Microscopías Avanzadas (LMA), University of Zaragoza, Zaragoza, Spain

Cesare Orlandi
Department of Pharmacology and Physiology, University of Rochester Medical Center,
Rochester, NY, United States

Barnali Paul
Center for Clinical Pharmacology, University of Health Sciences & Pharmacy at St Louis and
Washington University School of Medicine, St Louis, MO, United States

Berkay Selçuk
Molecular Biology, Genetics and Bioengineering Program, Faculty of Engineering and
Natural Sciences, Sabanci University, Istanbul, Turkey

Sashrik Sribhashyam
Center for Clinical Pharmacology, University of Health Sciences & Pharmacy at St Louis and
Washington University School of Medicine, St Louis, MO, United States

Doreen Thor
Rudolf Schönheimer Institute of Biochemistry, Medical Faculty, Leipzig University,
Leipzig, Germany

Alejandra Tomas
Section of Cell Biology and Functional Genomics, Department of Metabolism, Digestion
and Reproduction, Imperial College London, London, United Kingdom

Ayat Zagzoog
College of Pharmacy and Nutrition, University of Saskatchewan, Saskatoon, SK, Canada

Preface

G-protein-coupled receptors (GPCRs) are integral membrane proteins with a conserved seven-transmembrane architecture. There are more than 800 GPCRs in the human genome, and they are activated upon binding of a broad repertoire of stimuli. Their downstream signaling is critical for almost every cellular and physiological process, which also makes them one of the prime therapeutic targets. Over the past few years, several new paradigms of GPCR activation and signaling have emerged, and these developments have refined our understanding of the inner working and regulation of this versatile class of receptors. In this backdrop, we present Volumes 193 and 195 of *Progress in Molecular Biology and Translational Science*, focused on emerging directions in GPCR biology. The chapters included in the aforementioned volumes cover a broad repertoire of topics ranging from receptor activation and downstream signaling mechanisms to regulatory paradigms. In addition, the volumes also include chapters on the development of novels tools and approaches to probe the relatively unexplored aspects of receptor-ligand interactions and novel drug discovery. We sincerely hope that the topics covered in the two volumes will appeal to a large community of scientists engaged in GPCR research. I take this opportunity to express my sincere gratitude to all the contributors of the two volumes for taking the time to put together their chapters and also for their timely submission. I also thank our excellent production team for ensuring timely and efficient handling of the chapters; the two volumes would not have been possible without their fantastic effort. I hope that you find these two volumes useful, and I look forward to your comments and feedback.

ARUN K. SHUKLA, PhD
Department of Biological Sciences and Bioengineering
Indian Institute of technology, Kanpur 208016, India

> CHAPTER ONE

Adhesion G protein-coupled receptors—Structure and functions

Doreen Thor and Ines Liebscher*

Rudolf Schönheimer Institute of Biochemistry, Medical Faculty, Leipzig University, Leipzig, Germany
*Corresponding author: e-mail address: ines.liebscher@medizin.uni-leipzig.de

Contents

Abstract

Adhesion G protein-coupled receptors (aGPCRs) are an ancient class of receptors that represent some of the largest transmembrane-integrated proteins in humans. First recognized as surface markers on immune cells, it took more than a decade to appreciate their 7-transmembrane structure, which is reminiscent of GPCRs. Roughly 30 years went by before the first functional proof of an interaction with a G protein was published. Besides classic features of GPCRs (extracellular N terminus, 7-transmembrane region, intracellular C terminus), aGPCRs display a distinct N-terminal structure, which harbors the highly conserved GPCR autoproteolysis-inducing (GAIN) domain with the GPCR proteolysis site (GPS) in addition to several functional domains. Several human diseases have been associated with variants of aGPCRs and subsequent animal models have been established to investigate these phenotypes. Much progress has been made in recent years to decipher the structure and functions of these receptors. This chapter gives an overview of our current understanding with respect to the molecular structural patterns governing aGPCR activation and the contribution of these giant molecules to the development of pathologies.

Abbreviations

aGPCR/ADGR	adhesion G protein-coupled receptor
CTF	C-terminal fragment
GAIN	GPCR autoproteolysis-inducing
GPS	GPCR proteolysis site
NTF	N-terminal fragment
TM	transmembrane

Progress in Molecular Biology and Translational Science, Volume 195
ISSN 1877-1173
https://doi.org/10.1016/bs.pmbts.2022.06.009
1

1. An introduction to the class of adhesion G protein-coupled receptors

Adhesion-type GPCRs are a group of highly conserved and evolutionary old receptors, which are present in most animals and also in some primitive unicellular eukaryotes.[1] In humans, 33 representatives have been identified, yet numerous splice variants contribute to an even bigger variety with potential implications on signal induction.[2–5] A few years ago a new nomenclature has been approved by the IUPHAR, which assigns each of these receptors to one out of nine established families (I-IX): ADGRL (I), ADGRE (II), ADGRA (III), ADGRC (IV), ADGRD (V), ADGRF (VI), ADGRB (VII), ADGRG (VIII), and ADGRV (IX).[6]

With up to 6500 amino acids, aGPCRs are the largest membrane proteins known in humans. Several functional domains are located within the N terminus and have been shown to convey cell-cell or cell-matrix interaction, leading to a proposed dual role in cell adhesion and cell signal transduction.[7] The remarkable functions of the N terminus are most likely the reason why the first description of an aGPCR in 1981 missed the fact that it is a receptor molecule. The murine form of EMR1/ADGRE1 was recognized as specific surface marker of macrophages[8] and referred to as F4/80. It is still frequently used to identify macrophage populations. It took another 15 years to understand that this molecule also contains a 7-transmembrane (7TM) region, typical of GPCRs[9] and to remark the high similarity to another leucocyte surface molecule, CD97/ADGRE5, that has been cloned and identified as member of the secretin-like receptor family just prior to this observation.[10] The 1990's and early 2000's saw the identification of several more representatives[11–18] of what would be classified as the adhesion family of GPCRs within the 2003 established GRAFS system.[19]

In contrast, the latrophilin-type aGPCRs were first recognized for their ability to bind alpha latrotoxin, the main component of the black widow spider venom just about the same time as work on F4/80 was initially presented.[20] Subsequent studies suggested GPCR-type signal transduction of this molecule as binding of latrotoxin leads to depolarization and calcium influx in rat pheochromocytoma cells in combination with the accumulation of inositol phosphates.[21] First attempts to purify the receptor of interest were conducted in 1985,[22] which gave an idea of the approximate size of the protein. A more detailed description of the alpha-latrotoxin receptor portions, namely alpha, beta, and gamma, as well as their size and capability for each of them to bind latrotoxin were identified a few years later.[23] This description could be considered the first evidence for the characteristic structural features that are used today to describe aGPCRs.

During the first twenty years of aGPCR research members of this receptor class were identified and groundbreaking discoveries of their structural and functional principles made. The following 20 years were then dedicated to the refining these findings and the ever-growing availability of sophisticated research methods established several new signaling paradigms. This chapter aims to give a brief overview of our current knowledge of aGPCR structure as well as their molecular and physiological functions with pathophysiological implications.

2. Structural components of aGPCRs

Adhesion-type GPCRs display classic hallmarks of GPCRs like an extracellular N terminus, seven helices forming the transmembrane domain and an intracellular C terminus (Fig. 1). Only recently, the first cryo-electron microscopy (cryo-EM) structure of GPR97–Go complexes in the active state has been reported.[24] A characteristic feature of aGPCRs is the existence of a GPCR proteolysis site (GPS) in which the receptor

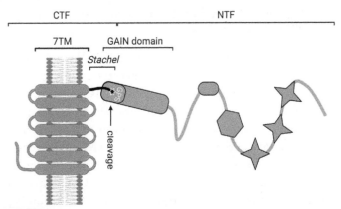

Fig. 1 General features of an Adhesion GPCR. Upon cleavage (indicated with a star) at a highly conserved GPCR proteolysis site (GPS) within the GPCR autoproteolysis-inducing (GAIN) domain, aGPCRs are divided into two functional subunits: the N-terminal fragment (NTF) and the C-terminal fragment (CTF). While the CTF harbors most of the general GPCR features like a 7TM region and the C terminus as well as the agonistic *Stachel* sequence, the NTF contains often several adhesive domains (displayed as stars, hexagon, and oval). *Figure created with biorender.*

is auto-proteolytically cleaved into an N-terminal and a C-terminal fragment (NTF and CTF, respectively)[25](Fig. 1). This cleavage event is based on a catalytic triade at a conserved cleavage motif (HL/I↓S/T) first described for EMR2/ADGRE2.[26,27] The GPS turned out to be the most C-terminal part of a larger highly conserved GPCR autoproteolysis-inducing (GAIN) domain that is both essential and sufficient to induce cleavage[28] (Fig. 1). The majority of aGPCRs harbor several additional N-terminal domains, which are characteristic for each family. While the initial assignment of these domains was based on comparative predictions, recent cryo-EM studies revealed the actual structural characteristics of several N-terminal domains in conjunction with the corresponding GAIN domains. For example, the Hormone Receptor (HormR) domains of LPHN1/ADGRL1, BAI3/ADGRB3,[28] and GPR126/ADGRG6[29] were solved. Other previously assumed and now structurally verified domains are the Complement C1r/C1s, Uegf, Bmp1 (CUB), and Pentraxin (PTX) domains of GPR126/ADGRG6.[29] Interestingly, these studies identified additional unexpected domains such as the Pentraxin/Laminin/neurexin/sex-hormone-binding-globulin-Like (PLL) domain in GPR56/ADGRG1[4] and a Sperm protein, Enterokinase and Agrin (SEA) domain in GPR126/ADGRG6.[29]

3. Structure follows function—The multiple facets of aGPCR activation

The large multidomain structures of aGPCRs are evolutionary conserved, which underlines their importance for proper receptor function.[30–32] The role of the 7TM domain of aGPCRs is equivalent to that of other GPCR classes; it mediates the intracellular signal transduction through G proteins, arrestins, and other intracellular adapters such as disheveled or ELMO (summarized in[33]) (Fig. 2A). The 7TM domain encompasses further the binding pocket for agonistic peptides and other soluble agonists such as glucocorticoids[24,34] or Wnt.[35–38] The N-terminal domains (including the GAIN domain) are interaction hubs with the surrounding extracellular matrix (ECM), cell surface antigens, or antibodies[33] (Fig. 2B). A direct modulation of the receptor's activity through the N-terminal binding partners was seen for some but not all of them. Among the activating ligands are collagen III, which could be shown to elicit downstream signaling in GPR56/ADGRG1-transfected cells[39] as well as collagen IV and laminin-211, which modulate GPR126/ADGRG6 activity.[40,41]

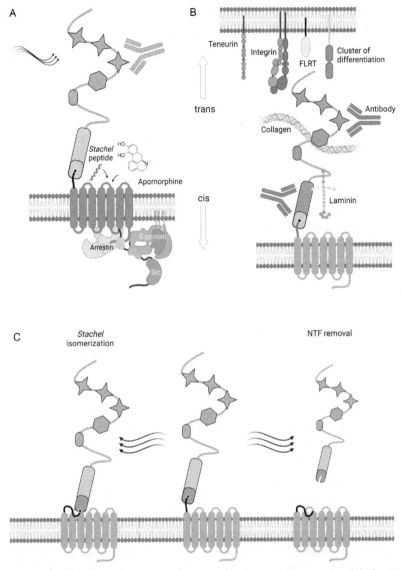

Fig. 2 Structural components of aGPCRs and their functional properties. (A) The 7TM domain provides a binding pocket for soluble ligands, which modulate the interaction of the receptor with its intracellular signaling molecules. (B) Domains of the N terminus interact with the surrounding ECM and adjacent cells. This interaction can be used to transmit a *trans* signal to the respective binding partner or a *cis* signal mediated by the respective receptor. (C) The role of cleavage for receptor (*Stachel*)-mediated mechanoactivation is still under debate and differs among aGPCR representatives. In a cleavage-independent model (left side) the agonistic sequence is already prebound and isomerizes under force application. Here, cleavage can serve as a structural prerequisite allowing for more flexibility. In the cleavage-dependent model (right side), mechanical forces induce the dissociation of the NTF, which exposes the *Stachel* sequence. In this model, cleavage is a prerequisite for receptor division. *Figure created with biorender.*

N-terminal antibodies can also activate aGPCRs as directly shown for GPR56/ADGRG1[42] and EMR2/ADGRE2.[43,44] Interestingly, the small molecule compound synaptamide also induces GPR110/ADGRF1-mediated cAMP production.[45] For the remaining ligand-receptor pairs one could speculate, if the right stimulus or read-out to determine receptor activation has not been applied yet. Alternatively, it could be possible that the physiological role of this interaction is to mediate signals independent from the classic intracellular signaling such as adherens or a so-called *trans* signal to stimulate the trans binding partner. There is mounting evidence that one aGPCR could mediate both, intracellular *cis* and extracellular *trans* signals[46–50] (Fig. 2B).

To achieve receptor activation through ECM molecules or surface antigens, one has to contemplate that long-lived, stable molecules with essentially no diffusivity do not represent classic characteristics of a receptor agonist. The fact that constitutive activity of an aGPCR can be achieved through the deletion of a large portion of the ectodomain, which serves as binding site for these ligands,[51–54] leaves only the explanation that the agonist is part of the receptor structure itself. This encrypted agonistic sequence, coined the *Stachel* sequence, has been located at the very N-terminal part of the receptor's CTF in 9 aGPCRs and derived synthetic peptides can be used to activate the receptors.[55–63] This appears to be a straightforward scenario: the N-terminal ligands bind and lead to the exposure of the agonistic sequence resulting in the active receptor conformation. The question, however, remains just how this binding event can induce changes in receptor conformation. Looking at the available crystal structures, it becomes apparent that there is a lot more to this seemingly simple story. Already the first crystal structures of the GAIN domains of LPHN1/ADGRL1 and BAI3/ADGRB3[28] showed that the hydrophobic *Stachel* sequence is deeply buried within this highly complex domain, raising the question of its possible exposure. However, only recently the cryo-EM structures of 7 aGPCR-CTFs were released, which show that the *Stachel* resides within the 7TM domain in this active receptor conformation.[64,65–67] Even as there is still no full-length structure available, which would provide us with the physiologic position of the *Stachel*, the triggers leading to its interaction with the binding pocket are highly sought after. With the *Stachel* sequence starting right at the cleavage site,[55] a role for this autoproteolytic event seems to be considered self-evident as it may facilitate the dissociation of the NTF from the CTF thereby releasing the tethered agonist.[55,56]

Already before the discovery of the *Stachel* sequence the importance of this cleavage had been highly debated. Mutations introduced to suppress the proteolytic procession appeared to significantly affect receptor trafficking and function for some representatives such as Latrophilin/Cirl,[25] GPR126/DREG/ADGRG6[68] or CD97/ADGRE5,[69] sparking speculations on a split receptor model, in which the NTF of one adhesion GPCR can functionally interact with the CTF of another.[70] This theory was later rejected through *in vivo* and *in vitro* experiments that report a much less prominent role for this cleavage event by rescuing a lethal phenotype in worm deficient for lat-1 with a cleavage-impaired mutant receptor version.[47] Furthermore, signal transduction studies demonstrated similar activity levels for receptors with a hampered cleavage motif and the respective wildtypes.[71] In addition to that some aGPCRs are not cleaved even though they carry the conserved cleavage motif.[58] As such the role for cleavage remains undecided and also the importance of the *Stachel* sequence has been challenged for some aGPCRs.[4,72]

Nevertheless, spontaneous NTF release was reported for GPR133/ADGRD1[73] while NTF dissociation of CD97/ADGRE5 has been observed upon interaction with its ligand CD55 under shear stress[74] or interaction with a monoclonal antibody.[75] GPR56/ADGRG1 interaction with heparin reduced shedding[76] but shear stress also led to a collagen-mediated release of the NTF.[77]

Independent from its effect on NTF/CTF dissociation the addition of mechanical forces appears to be necessary for some aGPCRs to induce activation with[41,77] and without[58,78] the interaction with a binding partner. These forces such as vibration, shaking, or shear stress provide the required momentum to induce an acute signal, which would not be elicited through ECM molecules or surface antigens alone. The exact molecular mechanism behind the mechano-activation is still under debate and will probably differ between aGPCRs representatives. Removal of the NTF and isomerization of the tethered agonist are both conceivable scenarios (Fig. 2C). Their main functional difference might be found in the persistence of the signal. An isomerization under mechanical forces allows for a reversible mode of action. In contrast the dissociation of the NTF is a one-and-done scenario.[56]

In conclusion, also for aGPCRs structure follows function, but with the enormous repertoire of transcripts that can arise from one aGPCR gene alone we are far from seeing all the naturally occurring structures and functions. At this state, we are collecting evidence for possible scenarios but only the future will provide us with a comprehensive picture.

3.1 Physiology and disease association

With their broad expression profile and functions in cell-cell and cell-matrix interactions[6,79,80] it is conceivable that aGPCRs have important physiological functions in humans and dysfunction might be the cause of severe diseases. Further evidence for the functional relevance can be deduced from different knock-out mouse models.

To search for variant-disease-associations within the aGPCR class in humans, we have used the DisGeNET database[81] and found 241 single nucleotide polymorphisms (SNPs) implicated in disease formation. Almost 50% of these SNPs are intron variants whose functional consequence is hard to predict. We, therefore, focused on SNPs causing 40 missense mutations, 27 premature stop codons, and 17 frameshift causing insertion/deletions (Table 1) as these directly interfere with receptor structure. Such mutations have been found for LPHN3/ADGRL3, LPHN4/ADGRL4, EMR2/ADGRE2, CELSR1/ADGRC1, CELSR2/ADGRC2, GPR115/ADGRF4, BAI2/ADGRB2, BAI3/ADGRB3, GPR56/ADGRG1, GPR64/ADGRG2, GPR97/ADGRG3, GPR126/ADGRG6, and VLGR/ADGRV1. For LPHN1/ADGRL1, EMR4/ADGRE4, GPR144/ADGRD2, GPR110/ADGRF1, GPR113/ADGRF3, GPR112/ADGRG5, GPR114/ADGRG5, and GPR128/ADGRG7 DisGeNET does not list any variants having a disease association. In the following we will summarize these associations and—if available—evidence from cell culture and/or animal models strengthening the hypotheses.

LPHN3/ADGRL3 has a described role in brain development coordinating synaptogenesis.[82] Genetic variants have a strong correlation with the occurrence of Attention-deficit/hyperactivity disorder.[83,84] Knock-out animal models confirm these findings as they develop hyperactivity.[85,86] Furthermore, Lphn3/Adgrl3 knock-out in mice changes expression of GABA and Serotonin receptors necessary for proper neurotransmission.[87]

GWAS analysis linked variants of EMR2/ADGRE2 with major depressive disorder[88] and vibratory urticaria.[89] The missense mutation for the latter, found within the GPS motif, represents a gain-of-function mutant. The exchange of C492Y appears to enhance the dissociation of NTF and CTF under vibration, which could lead to an enhanced activation of the receptor under mechanical stress resulting in increased mast cell degranulation.

The mutations I2107V and T2268A in CELSR1/ADGRC1 were identified as risk variants for ischemic stroke in different populations.[90–92] It was

Table 1 Variants associated with human diseases.

Gene	Variant	Alleles	Type of mutation	Disease
ADGRL3	rs734644	T/A; T/C	Missense	Attention deficit hyperactivity disorder
ADGRL4	rs41313381	C/A; C/T	Missense	RDW - Red blood cell distribution width result; White Blood Cell Count procedure; Eosinophil count procedure; Red cell distribution width determination
ADGRE2	rs112610420	T/G	Missense	Major Depressive Disorder
	rs199718602	C/T	Missense	Vibratory urticaria
ADGRC1	rs12170597	G/A	Missense	Susceptibility to neural tube defects
	rs1569226110	T/–	Frameshift	Milroy Disease
	rs1569227576	C/A	Premature stop codons	Milroy Disease
	rs199688538	C/T	Missense	Susceptibility to neural tube defects
	rs4044210	T/A; T/C	Missense	Ischemic stroke; Cerebrovascular accident; Large-artery atherosclerosis (embolus/thrombosis)
	rs6007897	T/C	Missense	Ischemic stroke; Cerebrovascular accident; Large-artery atherosclerosis (embolus/thrombosis)
	rs6008777	G/A	Missense	Susceptibility to neural tube defects
	rs61741871	G/C	Missense	Susceptibility to neural tube defects
	rs78201015	–/CA	Frameshift	Susceptibility to neural tube defects
	rs78201016	CA/–	Frameshift	Susceptibility to neural tube defects

Continued

Table 1 Variants associated with human diseases.—cont'd

Gene	Variant	Alleles	Type of mutation	Disease
ADGRC2	rs561330579	C/G; C/T	Missense	Global developmental delay; Poor school performance
ADGRF4	rs572583506	C/T	Missense	Malignant neoplasm of breast
ADGRB2	rs778361520	G/A	Missense	Progressive spastic paraparesis; Urinary Incontinence; Fecal Incontinence; Gait abnormality; Gait abnormality; Gait abnormality; Cachexia; Increased CSF protein; Abnormality of somatosensory evoked potentials; Nystagmus; Neurologic Symptoms
ADGRB3	rs1562005199	G/A	Missense	Short stature
	rs1562137453	T/C	Missense	Short stature
ADGRG1	rs121908462	C/A; C/T	Missense	Bilateral frontoparietal polymicrogyria
	rs121908463	T/A	Missense	Bilateral frontoparietal polymicrogyria
	rs121908464	C/T	Missense	Bilateral frontoparietal polymicrogyria
	rs121908465	G/C	Missense	Bilateral frontoparietal polymicrogyria
	rs121908466	A/G	Missense	Bilateral frontoparietal polymicrogyria
	rs14284762	A/G	Missense	Dysmorphic features; Muscle hypotonia; Movement Disorders
	rs146278035	C/T	Premature stop codons	Bilateral frontoparietal polymicrogyria
	rs1567782714	C/T	Premature stop codons	Bilateral frontoparietal polymicrogyria
	rs1567815105	–/T	Frameshift	Moderate intellectual disability, Low Vision, Global developmental delay, Generalized seizures

	rs532188689	G/C	Missense	Bilateral frontoparietal polymicrogyria
	rs556518689	G/A	Missense	Bilateral frontoparietal polymicrogyria
	rs587776625	CAGGACC/–	Frameshift	Bilateral frontoparietal polymicrogyria
	rs587783652	C/G; C/T	Premature stop codons	Bilateral frontoparietal polymicrogyria
	rs587783653	T/C	Missense	Bilateral frontoparietal polymicrogyria
	rs587783654	T/C	Missense	Bilateral frontoparietal polymicrogyria
	rs587783655	T/G	Premature stop codons	Bilateral frontoparietal polymicrogyria
	rs587783656	G/C	Missense	Bilateral frontoparietal polymicrogyria
	rs587783658	C/A; C/T	Missense	Bilateral frontoparietal polymicrogyria
	rs764367185	G/A	Missense	Bilateral frontoparietal polymicrogyria
	rs768441855	C/G; C/T	Premature stop codons	Bilateral frontoparietal polymicrogyria
	rs786204777	C/T	Premature stop codons	Bilateral frontoparietal polymicrogyria
	rs797045600	C/–	Frameshift	Bilateral frontoparietal polymicrogyria
	rs797045602	–/TT	Frameshift	Bilateral frontoparietal polymicrogyria
	rs942831911	T/G	Missense	Bilateral frontoparietal polymicrogyria
ADGRG2	rs774488954	A/–; A/AA	Frameshift	Infertility, Obstructive azoospermia
	rs879255538	A/–	Frameshift	Infertility, Obstructive azoospermia
	rs879255539	CACAG/TCT	Frameshift	Infertility, Obstructive azoospermia
ADGRG6	rs11155242	A/C	Missense	Pulmonary function
	rs17280293	A/G	Missense	Vital capacity; peak expiratory flow (procedure)

Continued

Table 1 Variants associated with human diseases.—cont'd

Gene	Variant	Alleles	Type of mutation	Disease
	rs536714306	G/A	Missense	Acute periodontitis; Aggressive Periodontitis
	rs749355583	C/G; C/T	Premature stop codons	Arthrogryposis
	rs793888524	–/G	Frameshift	Arthrogryposis
	rs793888525	T/A	Missense	Lethal congenital contracture syndrome 9; Arthrogryposis
ADGRV1	rs1057519383	G/A; G/T	Missense	Usher Syndrome, Type I
	rs1131691924	C/T	Missense	Usher syndrome, type 2C
	rs121909761	C/A	Premature stop codons	Familial febrile convulsions
	rs121909762	C/G; C/T	Premature stop codons	Usher Syndrome
	rs121909763	A/G	Missense	Usher syndrome, type 2C
	rs1554090072	C/G	Premature stop codons	Usher Syndrome
	rs1554117973	G/C	Missense	Usher Syndrome
	rs1561416879	C/G	Premature stop codons	Premature canities
	rs1561441451	G/T	Premature stop codons	Hearing impairment
	rs1561543496	G/T	Premature stop codons	Usher syndrome, type 2C
	rs1561740143	C/G	Premature stop codons	Usher Syndrome
	rs1561790371	G/T	Premature stop codons	Usher syndrome, type 2C
	rs1561843914	G/T	Premature stop codons	Usher syndrome, type 2C

rs200945405	G/A; G/T	Missense	Usher Syndrome
rs2247870	G/A	Missense	Schizophrenia
rs369793306	C/G; C/T	Premature stop codons	Usher syndrome, type 2C
rs373780305	C/T	Premature stop codons	Usher syndrome, type 2C
rs377650415	G/A; G/T	Premature stop codons	Usher syndrome, type 2C
rs527236131	C/T	Premature stop codons	Usher syndrome, type 2C
rs527236133	C/T	Premature stop codons	Usher syndrome, type 2C
rs747622607	C/T	Premature stop codons	Usher syndrome, type 2C
rs749956288	C/A; C/T	Premature stop codons	Usher Syndrome
rs756460900	A/G	Missense	Usher syndrome, type 2C
rs757696771	AT/–	Frameshift	Usher syndrome, type 2C
rs765574676	C/T	Missense	Hearing impairment
rs777309662	T/A	Premature stop codons	Usher Syndrome
rs796051863	–/AACA	Frameshift	Usher syndrome, type 2C
rs796051864	C/–	Frameshift	Usher syndrome, type 2C
rs796051866	AAGTGCTGAAATC/–	Frameshift	Usher syndrome, type 2C
rs796051867	AA/–	Frameshift	Usher syndrome, type 2C
rs878853348	TTCC/–	Frameshift	Retinal Dystrophies
rs886039893	G/T	Premature stop codons	Usher syndrome, type 2C

The database DisGeNET was searched for aGPCR variants associated with human diseases. Thereby, we focused on mutations changing the amino acid composition of the respected receptor by inducing missense mutations, frameshifts or premature stop codons.

speculated that I2107V within the hormone receptor domain interferes with ligand binding and, thereby, alters intracellular signaling, however, no functional analysis has been performed to verify this.[91] N- and C-terminal missense mutations of CELSR1/ADGRC1 were also described to cause craniorachischisis, the most severe form of neural tube defects.[93] Because all mutations resulted in a decrease in cell surface expression of CELSR1/ADGRC1 it is speculated that disruption of membrane localization is the disease cause.[93] Perinatal lethality due to neural tube defects have also been reported for *Celsr1/Adgrc1* knock-out mice[94] confirming CELSR1/ADGRC1 malfunction as cause for this phenotype. A truncated form of CELSR1/ADGRC1 was discovered in patients with hereditary lymphedema.[95] Although this variant was not functionally analyzed, *in vitro* and *in vivo* studies previously showed an involvement of CELSR1/ADGRC1 in lymphatic valve formation.[96] Interestingly, women are more likely to develop lymphedema from CELSR1/ADGRC1 loss-of-function than men.[97]

For BAI2/ADGRB2 a constitutive active mutation in the C-terminal domain changing arginine to tryptophan was identified in samples from patients suffering from neurological symptoms such as gait difficulties, severe weakness, spastic symptoms, and nystagmus.[98] To identify signaling properties, *in vitro* analyses were carried out using an N-terminal truncated version of BAI2/ADGRB2. Besides the increase in basal activity, these outlined that this mutation results in increased receptor expression. Furthermore, the wildtype form activated the pertussis toxin-insensitive Gz protein whereas the mutation led to a G protein-coupling shift towards the inhibitory G protein Gi.[98]

GPR56/ADGRG1 has been directly linked to a severe brain malformation called bilateral frontoparietal polymicrogyria as several mutations within this receptor were discovered in patients.[99–104] Characterization of those variants revealed that mutations can interfere with receptor autoproteolysis,[100] cell surface expression,[99] or ligand binding.[105] Furthermore, studies using *Gpr56/Adgrg1* knock-out mice demonstrated a myelination defect caused by a reduced number of oligodendrocytes.[106]

Knock-out of the x-linked *Gpr64/Adgrg2* in mouse models results in male infertility[107,108] with the loss of *Gpr64/Adgrg2* leading to reduced expression of epididymidis-specific proteins necessary for sperm maturation. Similarly, truncation mutants of GPR64/ADGRG2 in humans result in azoospermia and, subsequently, infertility.[109] Later on, it was shown that GPR64/ADGRG2 regulates pH homeostasis in the efferent ductules in a Gq and ß-arrestin-1-dependent manner.[110]

Several SNPs within GPR126/ADGRG6 have been associated with pulmonary dysfunction with two of them being missense mutations.[111,112] However, to our knowledge, the functional consequences of them have not been studied, yet. Furthermore, GPR126/ADGRG6 loss-of-function is implicated to cause Arthrogryposis Multiplex Congenita by interfering with proper myelination,[113] while another SNP was identified to be a risk factor for aggressive periodontitis.[114] The periodontitis-causing mutation is located within the 7TM region and abolishes cAMP signaling and PKA activation subsequently interfering with the expression of calcification-related genes.[114]

The very large G protein-coupled receptor (VLGR)/ADGRV1 was found while searching for the gene that causes audiogenic epilepsy in mice known as "frings" mice. The gene, originally named MASS1 for monogenic audiogenic seizure-susceptible,[115] was later on identified as a fragment of VLGR1/ADGRV1 spanning the N-terminal part of the putative ligand-binding domain.[116] It was only 2 years later when mutants in the same receptor were located in patients suffering from Usher syndrome type II, which is a congenital combination of loss of vision and hearing.[117] The hunt for other disease-associated variants within VLGR1/ADGRV1 was continued in the meantime, culminating in the identification of 53 different mutations in a next generation sequencing approach investigating a cohort of 44 patients suffering from Usher syndrome.[118] Besides this well studied link of VLGR1/ADGRV1 with Usher syndrome, GWAS associated another missense variant of VLGR1/ADGRV1 to schizophrenia.[119]

Besides the here reported variants it can be assumed that this receptor group contributes to several other (poly)genetic diseases since several GWAS studies link variants in intron, 3′ and 5′ prime sequences, or splice sites to human diseases. Further support for this belief also stems from receptor-deficient animal models (Fig. 3). Knock-out of aGPCRs causes embryonic death (*Lphn2/Adgrl2,*[120] *Gpr124/Adgra2,*[121] *Gpr126/Adgrg6,*[122] *Celsr2/Adgrc2*[123]) and ventilation failure (*Celsr3/Adgrc3*[124]). Knock-out animals that survive to adulthood can present with severe organ impairment like the accumulation of alveolar surfactant phospholipids (*Gpr116/Adgrf5*[125–127]) or myocardial hypertrophy when challenged with left ventricular pressure overload (*Eltd1/Adgrl4*[128]). Especially neuronal tissue requires intact aGPCR function to migrate and assume correct orientation (*Celsr1–3/Adgrc1–3*[129,130]), to recognize mechanosensory stimuli (Cirl/latrophilin[78]), to maintain spatial learning and memory (*Bai1/Adgrb1,*[131] *Bai3/Adgrb3*[132]), intact synapse formation (*Gpr110/Adgrf1*[133]) and for

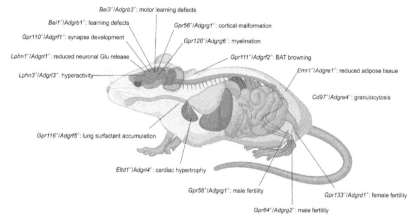

Fig. 3 The contribution of aGPCR malfunction to pathophysiological conditions has been evaluated in several constitutive and tissue- and cell-specific knock-out mouse models. Thereby, it was demonstrated that functionality of several aGPCRs is an essential requirement for neuron formation and viability. Furthermore, aGPCRs are involved in regulation of metabolic tissues, immune cell function, as well as fertility. *Mouse image taken from biorender.*

myelination as well as repair of peripheral axons (*Gpr126/Adgrg6*[134,135]). Reduction in male fertility has also been observed upon deletion of *Gpr56/Adgrg1*.[136] Additionally, several studies have shown an implication for this receptor family in immune defense,[137] metabolism,[138] and cancer.[139,140]

References

1. Kovacs P, Schöneberg T. The relevance of genomic signatures at adhesion GPCR loci in humans. *Handb Exp Pharmacol.* 2016;234:179–217. https://doi.org/10.1007/978-3-319-41523-9_9.
2. Knierim AB, Röthe J, Çakir MV, et al. Genetic basis of functional variability in adhesion G protein-coupled receptors. *Sci Rep.* 2019;9(1):11036. https://doi.org/10.1038/s41598-019-46265-x.
3. Boucard AA, Maxeiner S, Südhof TC. Latrophilins function as heterophilic cell-adhesion molecules by binding to teneurins: Regulation by alternative splicing. *J Biol Chem.* 2014;289(1):387–402. https://doi.org/10.1074/jbc.M113.504779.
4. Salzman GS, Ackerman SD, Ding C, et al. Structural basis for regulation of GPR56/ADGRG1 by its alternatively spliced extracellular domains. *Neuron.* 2016;91(6):1292–1304. https://doi.org/10.1016/j.neuron.2016.08.022.
5. Li T, Chiou B, Gilman CK, et al. A splicing isoform of GPR56 mediates microglial synaptic refinement via phosphatidylserine binding. *EMBO J.* 2020;39(16):e104136. https://doi.org/10.15252/embj.2019104136.
6. Hamann J, Aust G, Araç D, et al. International Union of Basic and Clinical Pharmacology. XCIV. Adhesion G protein-coupled receptors. *Pharmacol Rev.* 2015;67(2):338–367. https://doi.org/10.1124/pr.114.009647.

7. Yona S, Lin H-H, Siu WO, Gordon S, Stacey M. Adhesion-GPCRs: Emerging roles for novel receptors. *Trends Biochem Sci.* 2008;33(10):491–500. https://doi.org/10. 1016/j.tibs.2008.07.005.

8. Hirsch S. Expression of the macrophage-specific antigen F4/80 during differentiation of mouse bone marrow cells in culture. *Journal of Experimental Medicine.* 1981;154 (3):713–725. https://doi.org/10.1084/jem.154.3.713.

9. McKnight AJ, Macfarlane AJ, Dri P, Turley L, Willis AC, Gordon S. Molecular cloning of F4/80, a murine macrophage-restricted cell surface glycoprotein with homology to the G-protein-linked transmembrane 7 hormone receptor family. *J Biol Chem.* 1996;271(1):486–489.

10. Hamann J, Eichler W, Hamann D, et al. Expression cloning and chromosomal mapping of the leukocyte activation antigen CD97, a new seven-span transmembrane molecule of the secretion receptor superfamily with an unusual extracellular domain. *J Immunol.* 1995;155(4):1942–1950.

11. Abe J, Suzuki H, Notoya M, Yamamoto T, Hirose S. Ig-Hepta, a novel member of the G protein-coupled Hepta-helical receptor (GPCR) family that has immunoglobulin-like repeats in a long N-terminal extracellular domain and defines a new subfamily of GPCRs. *J Biol Chem.* 1999;274(28):19957–19964. https://doi. org/10.1074/jbc.274.28.19957.

12. Stacey M, Lin HH, Hilyard KL, Gordon S, McKnight AJ. Human epidermal growth factor (EGF) module-containing mucin-like hormone receptor 3 is a new member of the EGF-TM7 family that recognizes a ligand on human macrophages and activated neutrophils. *J Biol Chem.* 2001;276(22):18863–18870. https://doi.org/ 10.1074/jbc.M101147200.

13. Lin HH, Stacey M, Hamann J, Gordon S, McKnight AJ. Human EMR2, a novel EGF-TM7 molecule on chromosome 19p13.1, is closely related to CD97. *Genomics.* 2000;67(2):188–200. https://doi.org/10.1006/geno.2000.6238.

14. Caminschi I, Lucas KM, O'Keeffe MA, et al. Molecular cloning of F4/80-like-receptor, a seven-span membrane protein expressed differentially by dendritic cell and monocyte-macrophage subpopulations. *J Immunol.* 2001;167(7):3570–3576.

15. Stacey M, Chang G-W, Sanos SL, et al. EMR4, a novel epidermal growth factor (EGF)-TM7 molecule up-regulated in activated mouse macrophages, binds to a putative cellular ligand on B lymphoma cell line A20. *J Biol Chem.* 2002;277 (32):29283–29293. https://doi.org/10.1074/jbc.M204306200.

16. Shiratsuchi T, Nishimori H, Ichise H, Nakamura Y, Tokino T. Cloning and characterization of BAI2 and BAI3, novel genes homologous to brain-specific angiogenesis inhibitor 1 (BAI1). *Cytogenet Cell Genet.* 1997;79(1–2):103–108.

17. Hadjantonakis AK, Formstone CJ, Little PF. mCelsr1 is an evolutionarily conserved seven-pass transmembrane receptor and is expressed during mouse embryonic development. *Mech Dev.* 1998;78(1–2):91–95.

18. Fredriksson R, Gloriam DEI, Höglund PJ, Lagerström MC, Schiöth HB. There exist at least 30 human G-protein-coupled receptors with long ser/Thr-rich N-termini. *Biochem Biophys Res Commun.* 2003;301(3):725–734.

19. Fredriksson R, Lagerstrom MC, Lundin L-G, Schioth HB. The G-protein-coupled receptors in the human genome form five main families. Phylogenetic analysis, paralogon groups, and fingerprints. *Mol Pharmacol.* 2003;63(6):1256–1272. https://doi. org/10.1124/mol.63.6.1256.

20. Meldolesi J. Studies on alpha-latrotoxin receptors in rat brain synaptosomes: Correlation between toxin binding and stimulation of transmitter release. *J Neurochem.* 1982;38(6):1559–1569.

21. Vicentini LM, Meldolesi J. Alpha latrotoxin of black widow spider venom binds to a specific receptor coupled to phosphoinositide breakdown in PC12 cells. *Biochem Biophys Res Commun.* 1984;121(2):538–544.

22. Scheer H, Meldolesi J. Purification of the putative alpha-latrotoxin receptor from bovine synaptosomal membranes in an active binding form. *EMBO J.* 1985;4 (2):323–327.

23. Petrenko AG, Kovalenko VA, Shamotienko OG, et al. Isolation and properties of the alpha-latrotoxin receptor. *EMBO J.* 1990;9(6):2023–2027.

24. Ping Y-Q, Mao C, Xiao P, et al. Structures of the glucocorticoid-bound adhesion receptor GPR97-G(o) complex. *Nature.* 2021;589(7843):620–626. https://doi.org/ 10.1038/s41586-020-03083-w.

25. Krasnoperov V, Lu Y, Buryanovsky L, Neubert TA, Ichtchenko K, Petrenko AG. Post-translational proteolytic processing of the calcium-independent receptor of alpha-latrotoxin (CIRL), a natural chimera of the cell adhesion protein and the G protein-coupled receptor. Role of the G protein-coupled receptor proteolysis site (GPS) motif. *J Biol Chem.* 2002;277(48):46518–46526. https://doi.org/10.1074/jbc. M206415200.

26. Chang G-W, Stacey M, Kwakkenbos MJ, Hamann J, Gordon S, Lin H-H. Proteolytic cleavage of the EMR2 receptor requires both the extracellular stalk and the GPS motif. *FEBS Lett.* 2003;547(1–3):145–150.

27. Lin H-H, Chang G-W, Davies JQ, Stacey M, Harris J, Gordon S. Autocatalytic cleavage of the EMR2 receptor occurs at a conserved G protein-coupled receptor proteolytic site motif. *J Biol Chem.* 2004;279(30):31823–31832. https://doi.org/10.1074/jbc. M402974200.

28. Araç D, Boucard AA, Bolliger MF, et al. A novel evolutionarily conserved domain of cell-adhesion GPCRs mediates autoproteolysis. *EMBO J.* 2012;31(6):1364–1378. https://doi.org/10.1038/emboj.2012.26.

29. Leon K, Cunningham RL, Riback JA, et al. Structural basis for adhesion G protein-coupled receptor Gpr126 function. *Nat Commun.* 2020;11(1):194. https:// doi.org/10.1038/s41467-019-14040-1.

30. Krishnan A, Nijmeijer S, Graaf Cd, Schiöth HB. Classification, nomenclature, and structural aspects of adhesion GPCRs. *Handb Exp Pharmacol.* 2016;234:15–41. https://doi.org/10.1007/978-3-319-41523-9_2.

31. Scholz N, Langenhan T, Schöneberg T. Revisiting the classification of adhesion GPCRs. *Ann N Y Acad Sci.* 2019;1456(1):80–95. https://doi.org/10.1111/nyas.14192.

32. Wittlake A, Prömel S, Schöneberg T. The evolutionary history of vertebrate adhesion GPCRs and its implication on their classification. *Int J Mol Sci.* 2021;22(21). https://doi. org/10.3390/ijms222111803.

33. Langenhan T. Adhesion G protein-coupled receptors-candidate metabotropic mechanosensors and novel drug targets. *Basic Clin Pharmacol Toxicol.* 2020;126 (Suppl 6):5–16. https://doi.org/10.1111/bcpt.13223.

34. Gupte J, Swaminath G, Danao J, Tian H, Li Y, Wu X. Signaling property study of adhesion G-protein-coupled receptors. *FEBS Lett.* 2012;586(8):1214–1219. https:// doi.org/10.1016/j.febslet.2012.03.014.

35. Junge HJ. Ligand-selective Wnt receptor complexes in CNS blood vessels: RECK and GPR124 plugged in. *Neuron.* 2017;95(5):983–985. https://doi.org/10.1016/j.neuron. 2017.08.026.

36. Martin M, Vermeiren S, Bostaille N, et al. Engineered Wnt ligands enable blood-brain barrier repair in neurological disorders. *Science.* 2022;375(6582):eabm4459. https://doi. org/10.1126/science.abm4459.

37. Chang J, Mancuso MR, Maier C, et al. Gpr124 is essential for blood-brain barrier integrity in central nervous system disease. *Nat Med.* 2017;23(4):450–460. https:// doi.org/10.1038/nm.4309.

38. Posokhova E, Shukla A, Seaman S, et al. GPR124 functions as a WNT7-specific coactivator of canonical β-catenin signaling. *Cell Rep.* 2015;10(2):123–130. https:// doi.org/10.1016/j.celrep.2014.12.020.

39. Luo R, Jeong S-J, Yang A, et al. Mechanism for adhesion G protein-coupled receptor GPR56-mediated RhoA activation induced by collagen III stimulation. *PLoS One.* 2014;9(6):e100043. https://doi.org/10.1371/journal.pone.0100043.

40. Paavola KJ, Sidik H, Zuchero JB, Eckart M, Talbot WS. Type IV collagen is an activating ligand for the adhesion G protein-coupled receptor GPR126. *Sci Signal.* 2014;7(338):ra76. https://doi.org/10.1126/scisignal.2005347.

41. Petersen SC, Luo R, Liebscher I, et al. The adhesion GPCR GPR126 has distinct, domain-dependent functions in Schwann cell development mediated by interaction with laminin-211. *Neuron.* 2015;85(4):755–769. https://doi.org/10.1016/j.neuron.2014.12.057.

42. Iguchi T, Sakata K, Yoshizaki K, Tago K, Mizuno N, Itoh H. Orphan G protein-coupled receptor GPR56 regulates neural progenitor cell migration via a G alpha 12/13 and rho pathway. *J Biol Chem.* 2008;283(21):14469–14478. https://doi.org/10.1074/jbc.M708919200.

43. Yona S, Lin H-H, Dri P, et al. Ligation of the adhesion-GPCR EMR2 regulates human neutrophil function. *FASEB J.* 2008;22(3):741–751. https://doi.org/10.1096/fj.07-9435com.

44. Bhudia N, Desai S, King N, et al. G protein-coupling of adhesion GPCRs ADGRE2/EMR2 and ADGRE5/CD97, and activation of G protein Signalling by an anti-EMR2 antibody. *Sci Rep.* 2020;10(1):1004. https://doi.org/10.1038/s41598-020-57989-6.

45. Huang BX, Hu X, Kwon H-S, et al. Synaptamide activates the adhesion GPCR GPR110 (ADGRF1) through GAIN domain binding. *Commun Biol.* 2020;3(1):109. https://doi.org/10.1038/s42003-020-0831-6.

46. Matúš D, Post WB, Horn S, Schöneberg T, Prömel S. Latrophilin-1 drives neuron morphogenesis and shapes chemo- and mechanosensation-dependent behavior in C. elegans via a trans function. *Biochem Biophys Res Commun.* 2022;589:152–158. https://doi.org/10.1016/j.bbrc.2021.12.006.

47. Prömel S, Frickenhaus M, Hughes S, et al. The GPS motif is a molecular switch for bimodal activities of adhesion class G protein-coupled receptors. *Cell Rep.* 2012;2(2):321–331. https://doi.org/10.1016/j.celrep.2012.06.015.

48. Patra C, van Amerongen MJ, Ghosh S, et al. Organ-specific function of adhesion G protein-coupled receptor GPR126 is domain-dependent. *Proc Natl Acad Sci U S A.* 2013;110(42):16898–16903. https://doi.org/10.1073/pnas.1304837110.

49. Ward Y, Lake R, Faraji F, et al. Platelets promote metastasis via binding tumor CD97 leading to bidirectional signaling that coordinates Transendothelial migration. *Cell Rep.* 2018;23(3):808–822. https://doi.org/10.1016/j.celrep.2018.03.092.

50. Tu Y-K, Duman JG, Tolias KF. The adhesion-GPCR BAI1 promotes excitatory synaptogenesis by coordinating bidirectional trans-synaptic signaling. *J Neurosci.* 2018;38(39):8388–8406. https://doi.org/10.1523/JNEUROSCI.3461-17.2018.

51. Okajima D, Kudo G, Yokota H. Brain-specific angiogenesis inhibitor 2 (BAI2) may be activated by proteolytic processing. *J Recept Signal Transduct Res.* 2010;30(3):143–153. https://doi.org/10.3109/10799891003671139.

52. Yang L, Chen G, Mohanty S, et al. GPR56 regulates VEGF production and angiogenesis during melanoma progression. *Cancer Res.* 2011;71(16):5558–5568. https://doi.org/10.1158/0008-5472.CAN-10-4543.

53. Paavola KJ, Stephenson JR, Ritter SL, Alter SP, Hall RA. The N terminus of the adhesion G protein-coupled receptor GPR56 controls receptor signaling activity. *J Biol Chem.* 2011;286(33):28914–28921. https://doi.org/10.1074/jbc.M111.247973.

54. Ward Y, Lake R, Yin JJ, et al. LPA receptor heterodimerizes with CD97 to amplify LPA-initiated RHO-dependent signaling and invasion in prostate cancer cells. *Cancer Res.* 2011;71(23):7301–7311. https://doi.org/10.1158/0008-5472.CAN-11-2381.

55. Liebscher I, Schon J, Petersen SC, et al. A tethered agonist within the ectodomain activates the adhesion G protein-coupled receptors GPR126 and GPR133. *Cell Rep.* 2014;9(6):2018–2026. https://doi.org/10.1016/j.celrep.2014.11.036.

56. Stoveken HM, Hajduczok AG, Xu L, Tall GG. Adhesion G protein-coupled receptors are activated by exposure of a cryptic tethered agonist. *Proc Natl Acad Sci U S A.* 2015;112(19):6194–6199. https://doi.org/10.1073/pnas.1421785112.

57. Müller A, Winkler J, Fiedler F, et al. Oriented cell division in the C. elegans embryo is coordinated by G-protein signaling dependent on the adhesion GPCR LAT-1. *PLoS Genet.* 2015;11(10):e1005624. https://doi.org/10.1371/journal.pgen.1005624.

58. Wilde C, Fischer L, Lede V, et al. The constitutive activity of the adhesion GPCR GPR114/ADGRG5 is mediated by its tethered agonist. *FASEB J.* 2015. https://doi.org/10.1096/fj.15-276220.

59. Demberg LM, Rothemund S, Schoneberg T, Liebscher I. Identification of the tethered peptide agonist of the adhesion G protein-coupled receptor GPR64/ADGRG2. *Biochem Biophys Res Commun.* 2015;464(3):743–747. https://doi.org/10.1016/j.bbrc.2015.07.020.

60. Röthe J, Thor D, Winkler J, et al. Involvement of the adhesion GPCRs Latrophilins in the regulation of insulin release. *Cell Rep.* 2019;26(6):1573–1584.e5. https://doi.org/10.1016/j.celrep.2019.01.040.

61. Demberg LM, Winkler J, Wilde C, et al. Activation of adhesion G protein-coupled receptors: Agonist specificity of stachel sequence-derived peptides. *J Biol Chem.* 2017;292(11):4383–4394. https://doi.org/10.1074/jbc.M116.763656.

62. Balenga N, Azimzadeh P, Hogue JA, et al. Orphan adhesion GPCR GPR64/ADGRG2 is overexpressed in parathyroid tumors and attenuates calcium-sensing receptor-mediated signaling. *J Bone Miner Res.* 2017;32(3):654–666. https://doi.org/10.1002/jbmr.3023.

63. Brown K, Filuta A, Ludwig M-G, et al. Epithelial Gpr116 regulates pulmonary alveolar homeostasis via Gq/11 signaling. *JCI Insight.* 2017;2(11). https://doi.org/10.1172/jci.insight.93700.

64. Qu X, et al. Structural basis of tethered agonism of the adhesion GPCRs ADGRD1 and ADGRF1. *Nature.* 2022. https://doi.org/10.1038/s41586-022-04580-w.

65. Ping Y-Q, et al. Structural basis for the tethered peptide activation of adhesion GPCRs. *Nature.* 2022. https://doi.org/10.1038/s41586-022-04619-y.

66. Xiao P, et al. Tethered peptide activation mechanism of the adhesion GPCRs ADGRG2 and ADGRG4. *Nature.* 2022. https://doi.org/10.1038/s41586-022-04590-8.

67. Barros-Álvarez X, et al. The tethered peptide activation mechanism of adhesion GPCRs. *Nature.* 2022. https://doi.org/10.1038/s41586-022-04575-7.

68. Moriguchi T, Haraguchi K, Ueda N, Okada M, Furuya T, Akiyama T. DREG, a developmentally regulated G protein-coupled receptor containing two conserved proteolytic cleavage sites. *Genes Cells.* 2004;9(6):549–560. https://doi.org/10.1111/j.1356-9597.2004.00743.x.

69. Hsiao C-C, Chen H-Y, Chang G-W, Lin H-H. GPS autoproteolysis is required for CD97 to up-regulate the expression of N-cadherin that promotes homotypic cell-cell aggregation. *FEBS Lett.* 2011;585(2):313–318. https://doi.org/10.1016/j.febslet.2010.12.005.

70. Silva J-P, Lelianova V, Hopkins C, Volynski KE, Ushkaryov Y. Functional cross-interaction of the fragments produced by the cleavage of distinct adhesion G-protein-coupled receptors. *J Biol Chem.* 2009;284(10):6495–6506. https://doi.org/10.1074/jbc.M806979200.

71. Bohnekamp J, Schöneberg T. Cell adhesion receptor GPR133 couples to Gs protein. *J Biol Chem.* 2011;286(49):41912–41916. https://doi.org/10.1074/jbc.C111.265934.

72. Kishore A, Purcell RH, Nassiri-Toosi Z, Hall RA. Stalk-dependent and stalk-independent signaling by the adhesion G protein-coupled receptors GPR56 (ADGRG1) and BAI1 (ADGRB1). *J Biol Chem.* 2016;291(7):3385–3394. https://doi.org/10.1074/jbc.M115.689349.

73. Frenster JD, Stephan G, Ravn-Boess N, et al. Functional impact of intramolecular cleavage and dissociation of adhesion G protein-coupled receptor GPR133 (ADGRD1) on canonical signaling. *J Biol Chem.* 2021;296. https://doi.org/10.1016/j.jbc.2021.100798, 100798.

74. Karpus ON, Veninga H, Hoek RM, et al. Shear stress-dependent downregulation of the adhesion-G protein-coupled receptor CD97 on circulating leukocytes upon contact with its ligand CD55. *J Immunol.* 2013;190(7):3740–3748. https://doi.org/10.4049/jimmunol.1202192.

75. Groot DMd, Vogel G, Dulos J, et al. Therapeutic antibody targeting of CD97 in experimental arthritis: The role of antigen expression, shedding, and internalization on the pharmacokinetics of anti-CD97 monoclonal antibody 1B2. *J Immunol.* 2009;183(6):4127–4134. https://doi.org/10.4049/jimmunol.0901253.

76. Chiang N-Y, Chang G-W, Huang Y-S, et al. Heparin interacts with the adhesion GPCR GPR56, reduces receptor shedding, and promotes cell adhesion and motility. *J Cell Sci.* 2016;129(11):2156–2169. https://doi.org/10.1242/jcs.174458.

77. Yeung J, Adili R, Stringham EN, et al. GPR56/ADGRG1 is a platelet collagen-responsive GPCR and hemostatic sensor of shear force. *Proc Natl Acad Sci U S A.* 2020;117(45):28275–28286. https://doi.org/10.1073/pnas.2008921117.

78. Scholz N, Gehring J, Guan C, et al. The adhesion GPCR latrophilin/CIRL shapes mechanosensation. *Cell Rep.* 2015;11(6):866–874. https://doi.org/10.1016/j.celrep.2015.04.008.

79. Scholz N, Monk KR, Kittel RJ, Langenhan T. Adhesion GPCRs as a putative class of metabotropic Mechanosensors. *Handb Exp Pharmacol.* 2016;234:221–247. https://doi.org/10.1007/978-3-319-41523-9_10.

80. Liebscher I, Cevheroğlu O, Hsiao C-C, et al. A guide to adhesion GPCR research. *FEBS J.* 2021. https://doi.org/10.1111/febs.16258.

81. Piñero J, Ramírez-Anguita JM, Saüch-Pitarch J, et al. The DisGeNET knowledge platform for disease genomics: 2019 update. *Nucleic Acids Res.* 2020;48(D1):D845–D855. https://doi.org/10.1093/nar/gkz1021.

82. Sando R, Jiang X, Südhof TC. Latrophilin GPCRs direct synapse specificity by coincident binding of FLRTs and teneurins. *Science.* 2019;363(6429). https://doi.org/10.1126/science.aav7969.

83. Ribasés M, Ramos-Quiroga JA, Sánchez-Mora C, et al. Contribution of LPHN3 to the genetic susceptibility to ADHD in adulthood: A replication study. *Genes Brain Behav.* 2011;10(2):149–157. https://doi.org/10.1111/j.1601-183X.2010.00649.x.

84. Bruxel EM, Salatino-Oliveira A, Akutagava-Martins GC, et al. LPHN3 and attention-deficit/hyperactivity disorder: A susceptibility and pharmacogenetic study. *Genes Brain Behav.* 2015;14(5):419–427. https://doi.org/10.1111/gbb.12224.

85. Regan SL, Hufgard JR, Pitzer EM, et al. Knockout of latrophilin-3 in Sprague-Dawley rats causes hyperactivity, hyper-reactivity, under-response to amphetamine, and disrupted dopamine markers. *Neurobiol Dis.* 2019;130:104494. https://doi.org/10.1016/j.nbd.2019.104494.

86. Regan SL, Pitzer EM, Hufgard JR, Sugimoto C, Williams MT, Vorhees CV. A novel role for the ADHD risk gene latrophilin-3 in learning and memory in Lphn3 knockout rats. *Neurobiol Dis.* 2021;158:105456. https://doi.org/10.1016/j.nbd.2021.105456.

87. Orsini CA, Setlow B, DeJesus M, et al. Behavioral and transcriptomic profiling of mice null for Lphn3, a gene implicated in ADHD and addiction. *Mol Genet Genomic Med.* 2016;4(3):322–343. https://doi.org/10.1002/mgg3.207.

88. Wong M-L, Arcos-Burgos M, Liu S, et al. The PHF21B gene is associated with major depression and modulates the stress response. *Mol Psychiatry.* 2017;22(7):1015–1025. https://doi.org/10.1038/mp.2016.174.

89. Boyden SE, Desai A, Cruse G, et al. Vibratory urticaria associated with a missense variant in ADGRE2. *N Engl J Med.* 2016;374(7):656–663. https://doi.org/10.1056/NEJMoa1500611.

90. Yamada Y, Fuku N, Tanaka M, et al. Identification of CELSR1 as a susceptibility gene for ischemic stroke in Japanese individuals by a genome-wide association study. *Atherosclerosis.* 2009;207(1):144–149. https://doi.org/10.1016/j.atherosclerosis.2009.03.038.

91. Zhan Y-H, Lin Y, Tong S-J, et al. The CELSR1 polymorphisms rs6007897 and rs4044210 are associated with ischaemic stroke in Chinese Han population. *Ann Hum Biol.* 2015;42(1):26–30. https://doi.org/10.3109/03014460.2014.944214.

92. Gouveia LO, Sobral J, Vicente AM, Ferro JM, Oliveira SA. Replication of the CELSR1 association with ischemic stroke in a Portuguese case-control cohort. *Atherosclerosis.* 2011;217(1):260–262. https://doi.org/10.1016/j.atherosclerosis.2011.03.022.

93. Robinson A, Escuin S, Doudney K, et al. Mutations in the planar cell polarity genes CELSR1 and SCRIB are associated with the severe neural tube defect craniorachischisis. *Hum Mutat.* 2012;33(2):440–447. https://doi.org/10.1002/humu.21662.

94. Curtin JA, Quint E, Tsipouri V, et al. Mutation of Celsr1 disrupts planar polarity of inner ear hair cells and causes severe neural tube defects in the mouse. *Curr Biol.* 2003;13(13):1129–1133. https://doi.org/10.1016/s0960-9822(03)00374-9.

95. Gonzalez-Garay ML, Aldrich MB, Rasmussen JC, et al. A novel mutation in CELSR1 is associated with hereditary lymphedema. *Vasc Cell.* 2016;8:1. https://doi.org/10.1186/s13221-016-0035-5.

96. Tatin F, Taddei A, Weston A, et al. Planar cell polarity protein Celsr1 regulates endothelial adherens junctions and directed cell rearrangements during valve morphogenesis. *Dev Cell.* 2013;26(1):31–44. https://doi.org/10.1016/j.devcel.2013.05.015.

97. Maltese PE, Michelini S, Ricci M, et al. Increasing evidence of hereditary lymphedema caused by CELSR1 loss-of-function variants. *Am J Med Genet A.* 2019;179(9):1718–1724. https://doi.org/10.1002/ajmg.a.61269.

98. Purcell RH, Toro C, Gahl WA, Hall RA. A disease-associated mutation in the adhesion GPCR BAI2 (ADGRB2) increases receptor signaling activity. *Hum Mutat.* 2017;38(12):1751–1760. https://doi.org/10.1002/humu.23336.

99. Chiang N-Y, Hsiao C-C, Huang Y-S, et al. Disease-associated GPR56 mutations cause bilateral frontoparietal polymicrogyria via multiple mechanisms. *J Biol Chem.* 2011;286(16):14215–14225. https://doi.org/10.1074/jbc.M110.183830.

100. Luo R, Yang HM, Jin Z, et al. A novel GPR56 mutation causes bilateral frontoparietal polymicrogyria. *Pediatr Neurol.* 2011;45(1):49–53. https://doi.org/10.1016/j.pediatrneurol.2011.02.004.

101. Kishore A, Hall RA. Disease-associated extracellular loop mutations in the adhesion G protein-coupled receptor G1 (ADGRG1; GPR56) differentially regulate downstream signaling. *J Biol Chem.* 2017;292(23):9711–9720. https://doi.org/10.1074/jbc.M117.780551.

102. Santos-Silva R, Passas A, Rocha C, et al. Bilateral frontoparietal polymicrogyria: A novel GPR56 mutation and an unusual phenotype. *Neuropediatrics.* 2015;46(2):134–138. https://doi.org/10.1055/s-0034-1399754.

103. Piao X, Hill RS, Bodell A, et al. G protein-coupled receptor-dependent development of human frontal cortex. *Science.* 2004;303(5666):2033–2036. https://doi.org/10.1126/science.1092780.

104. Piao X, Chang BS, Bodell A, et al. Genotype-phenotype analysis of human frontoparietal polymicrogyria syndromes. *Ann Neurol*. 2005;58(5):680–687. https://doi.org/10.1002/ana.20616.

105. Luo R, Jin Z, Deng Y, Strokes N, Piao X. Disease-associated mutations prevent GPR56-collagen III interaction. *PLoS One*. 2012;7(1):e29818. https://doi.org/10.1371/journal.pone.0029818.

106. Giera S, Deng Y, Luo R, et al. The adhesion G protein-coupled receptor GPR56 is a cell-autonomous regulator of oligodendrocyte development. *Nat Commun*. 2015;6:6121. https://doi.org/10.1038/ncomms7121.

107. Davies B, Behnen M, Cappallo-Obermann H, Spiess A-N, Theuring F, Kirchhoff C. Novel epididymis-specific mRNAs downregulated by HE6/Gpr64 receptor gene disruption. *Mol Reprod Dev*. 2007;74(5):539–553. https://doi.org/10.1002/mrd.20636.

108. Davies B, Baumann C, Kirchhoff C, et al. Targeted deletion of the epididymal receptor HE6 results in fluid dysregulation and male infertility. *Mol Cell Biol*. 2004;24(19):8642–8648. https://doi.org/10.1128/MCB.24.19.8642-8648.2004.

109. Patat O, Pagin A, Siegfried A, et al. Truncating mutations in the adhesion G protein-coupled receptor G2 gene ADGRG2 cause an X-linked congenital bilateral absence of vas deferens. *Am J Hum Genet*. 2016;99(2):437–442. https://doi.org/10.1016/j.ajhg.2016.06.012.

110. Zhang D-L, Sun Y-J, Ma M-L, et al. Gq activity- and β-arrestin-1 scaffolding-mediated ADGRG2/CFTR coupling are required for male fertility. *Elife*. 2018;7. https://doi.org/10.7554/eLife.33432.

111. Hancock DB, Eijgelsheim M, Wilk JB, et al. Meta-analyses of genome-wide association studies identify multiple loci associated with pulmonary function. *Nat Genet*. 2010;42(1):45–52. https://doi.org/10.1038/ng.500.

112. Shrine N, Guyatt AL, Erzurumluoglu AM, et al. New genetic signals for lung function highlight pathways and chronic obstructive pulmonary disease associations across multiple ancestries. *Nat Genet*. 2019;51(3):481–493. https://doi.org/10.1038/s41588-018-0321-7.

113. Ravenscroft G, Nolent F, Rajagopalan S, et al. Mutations of GPR126 are responsible for severe arthrogryposis multiplex congenita. *Am J Hum Genet*. 2015;96(6):955–961. https://doi.org/10.1016/j.ajhg.2015.04.014.

114. Kitagaki J, Miyauchi S, Asano Y, et al. A putative Association of a Single Nucleotide Polymorphism in GPR126 with aggressive periodontitis in a Japanese population. *PLoS One*. 2016;11(8):e0160765. https://doi.org/10.1371/journal.pone.0160765.

115. Skradski SL, White HS, Ptacek LJ. Genetic mapping of a locus (mass1) causing audiogenic seizures in mice. *Genomics*. 1998;49(2):188–192. https://doi.org/10.1006/geno.1998.5229.

116. McMillan DR, Kayes-Wandover KM, Richardson JA, White PC. Very large G protein-coupled receptor-1, the largest known cell surface protein, is highly expressed in the developing central nervous system. *J Biol Chem*. 2002;277(1):785–792. https://doi.org/10.1074/jbc.M108929200.

117. Weston MD, Luijendijk MWJ, Humphrey KD, Moller C, Kimberling WJ. Mutations in the VLGR1 gene implicate G-protein signaling in the pathogenesis of usher syndrome type II. *Am J Hum Genet*. 2004;74(2):357–366. https://doi.org/10.1086/381685.

118. Aparisi MJ, Aller E, Fuster-García C, et al. Targeted next generation sequencing for molecular diagnosis of usher syndrome. *Orphanet J Rare Dis*. 2014;9:168. https://doi.org/10.1186/s13023-014-0168-7.

119. Li Z, Chen J, Yu H, et al. Genome-wide association analysis identifies 30 new susceptibility loci for schizophrenia. *Nat Genet*. 2017;49(11):1576–1583. https://doi.org/10.1038/ng.3973.

120. Anderson GR, Maxeiner S, Sando R, Tsetsenis T, Malenka RC, Südhof TC. Postsynaptic adhesion GPCR latrophilin-2 mediates target recognition in entorhinal-hippocampal synapse assembly. *J Cell Biol.* 2017;216(11):3831–3846. https://doi.org/10.1083/jcb.201703042.
121. Kuhnert F, Mancuso MR, Shamloo A, et al. Essential regulation of CNS angiogenesis by the orphan G protein-coupled receptor GPR124. *Science.* 2010;330(6006): 985–989. https://doi.org/10.1126/science.1196554.
122. Waller-Evans H, Prömel S, Langenhan T, et al. The orphan adhesion-GPCR GPR126 is required for embryonic development in the mouse. *PLoS One.* 2010;5(11): e14047. https://doi.org/10.1371/journal.pone.0014047.
123. Tissir F, Qu Y, Montcouquiol M, et al. Lack of cadherins Celsr2 and Celsr3 impairs ependymal ciliogenesis, leading to fatal hydrocephalus. *Nat Neurosci.* 2010;13 (6):700–707. https://doi.org/10.1038/nn.2555.
124. Tissir F, Bar I, Jossin Y, Backer Od, Goffinet AM. Protocadherin Celsr3 is crucial in axonal tract development. *Nat Neurosci.* 2005;8(4):451–457. https://doi.org/10.1038/nn1428.
125. Bridges JP, Ludwig M-G, Mueller M, et al. Orphan G protein-coupled receptor GPR116 regulates pulmonary surfactant pool size. *Am J Respir Cell Mol Biol.* 2013;49(3):348–357. https://doi.org/10.1165/rcmb.2012-0439OC.
126. Yang MY, Hilton MB, Seaman S, et al. Essential regulation of lung surfactant homeostasis by the orphan G protein-coupled receptor GPR116. *Cell Rep.* 2013;3 (5):1457–1464. https://doi.org/10.1016/j.celrep.2013.04.019.
127. Fukuzawa T, Ishida J, Kato A, et al. Lung surfactant levels are regulated by Ig-Hepta/GPR116 by monitoring surfactant protein D. *PLoS One.* 2013;8(7):e69451. https://doi.org/10.1371/journal.pone.0069451.
128. Xiao J, Jiang H, Zhang R, et al. Augmented cardiac hypertrophy in response to pressure overload in mice lacking ELTD1. *PLoS One.* 2012;7(5):e35779. https://doi.org/10.1371/journal.pone.0035779.
129. Boutin C, Labedan P, Dimidschstein J, et al. A dual role for planar cell polarity genes in ciliated cells. *Proc Natl Acad Sci U S A.* 2014;111(30):E3129–E3138. https://doi.org/10.1073/pnas.1404988111.
130. Qu Y, Glasco DM, Zhou L, et al. Atypical cadherins Celsr1-3 differentially regulate migration of facial branchiomotor neurons in mice. *J Neurosci.* 2010;30 (28):9392–9401. https://doi.org/10.1523/JNEUROSCI.0124-10.2010.
131. Zhu D, Li C, Swanson AM, et al. BAI1 regulates spatial learning and synaptic plasticity in the hippocampus. *J Clin Invest.* 2015;125(4):1497–1508. https://doi.org/10.1172/JCI74603.
132. Kakegawa W, Mitakidis N, Miura E, et al. Anterograde C1ql1 signaling is required in order to determine and maintain a single-winner climbing fiber in the mouse cerebellum. *Neuron.* 2015;85(2):316–329. https://doi.org/10.1016/j.neuron.2014.12.020.
133. Lee J-W, Huang BX, Kwon H, et al. Orphan GPR110 (ADGRF1) targeted by N-docosahexaenoylethanolamine in development of neurons and cognitive function. *Nat Commun.* 2016;7:13123. https://doi.org/10.1038/ncomms13123.
134. Monk KR, Naylor SG, Glenn TD, et al. A G protein-coupled receptor is essential for Schwann cells to initiate myelination. *Science.* 2009;325(5946):1402–1405. https://doi.org/10.1126/science.1173474.
135. Mogha A, Harty BL, Carlin D, et al. Gpr126/Adgrg6 has Schwann cell autonomous and nonautonomous functions in peripheral nerve injury and repair. *J Neurosci.* 2016;36(49):12351–12367. https://doi.org/10.1523/JNEUROSCI.3854-15.2016.
136. Chen G, Yang L, Begum S, Xu L. GPR56 is essential for testis development and male fertility in mice. *Dev Dyn.* 2010;239(12):3358–3367. https://doi.org/10.1002/dvdy.22468.

137. Lin H-H, Hsiao C-C, Pabst C, Hébert J, Schöneberg T, Hamann J. Adhesion GPCRs in regulating immune responses and inflammation. *Adv Immunol.* 2017;136:163–201. https://doi.org/10.1016/bs.ai.2017.05.005.

138. Kaczmarek I, Suchý T, Prömel S, Schöneberg T, Liebscher I, Thor D. The relevance of adhesion G protein-coupled receptors in metabolic functions. *Biol Chem.* 2022;403 (2):195–209. https://doi.org/10.1515/hsz-2021-0146.

139. Lin H-H. Adhesion family of G protein-coupled receptors and cancer. *Chang Gung Med J.* 2012;35(1):15–27.

140. Gad AA, Balenga N. The emerging role of adhesion GPCRs in Cancer. *ACS Pharmacol Transl Sci.* 2020;3(1):29–42. https://doi.org/10.1021/acsptsci.9b00093.

CHAPTER TWO

Molecular and cellular mechanisms underlying brain region-specific endocannabinoid system modulation by estradiol across the rodent estrus cycle

Hye Ji J. Kim[a], Ayat Zagzoog[a], Tallan Black[a], Sarah L. Baccetto[a], and Robert B. Laprairie[a,b,*]

[a]College of Pharmacy and Nutrition, University of Saskatchewan, Saskatoon, SK, Canada
[b]Department of Pharmacology, College of Medicine, Dalhousie University, Halifax, NS, Canada
*Corresponding author: e-mail address: robert.laprairie@usask.ca

Contents

Abstract

Neurological crosstalk between the endocannabinoid and estrogen systems has been a growing topic of discussion over the last decade. Although the main estrogenic ligand, estradiol (E2), influences endocannabinoid signaling in both male and female animals, the latter experiences significant and rhythmic fluctuations in E2 as well as other sex hormones. This is referred to as the menstrual cycle in women and the estrus cycle in rodents such as mice and rats. Consisting of 4 distinct hormone-driven phases, the rodent estrus cycle modulates both endocannabinoid and exogenous cannabinoid signaling resulting in unique behavioral outcomes based on the cycle phase.

27

For example, cannabinoid receptor agonist-induced antinociception is greatest during proestrus and estrus, when circulating and brain levels of E2 are high, as compared to metestrus and diestrus when E2 concentrations are low. Pain processing occurs throughout the cerebral cortex and amygdala of the forebrain; periaqueductal grey of the midbrain; and medulla and spine of the hindbrain. As a result, past molecular investigations on these endocannabinoid-estrogen system interactions have focused on these specific brain regions. Here, we will bridge regional molecular trends with neurophysiological evidence of how plasma membrane estrogen receptor (ER) activation by E2 leads to postsynaptic endocannabinoid synthesis, retrograde signaling, and alterations in inhibitory neurotransmission. These signaling pathways depend on ER heterodimers, current knowledge of which will also be detailed in this review. Overall, the aim of this review article is to systematically summarize how the cannabinoid receptors and endocannabinoids change in expression and function in specific brain regions throughout the estrus cycle.

1. Introduction

The endocannabinoid system (ECS) is responsible for a wide range of central nervous system (CNS) functions including movement,[1] cognition, mood,[2] reward,[3] pain,[4] stress,[5] sleep,[6] appetite, and energy homeostasis.[7] The two primary G protein-coupled receptors (GPCR) of the ECS are: the type 1 cannabinoid receptor CB1R) and the type 2 cannabinoid receptor (CB2R).[8,9] The endogenous ligands or endocannabinoids that bind and activate these receptors are anandamide (AEA) and 2-arachidonoylglycerol (2-AG). There are also countless other exogenous plant cannabinoids (e.g., Δ^9-tetrahydrocannabinol and cannabidiol) and synthetic cannabinoids (e.g., the full cannabinoid receptor agonist CP55,940) that stimulate these receptor subtypes. Cannabinoids have distinct behavioral effects in males and females, and female animals have repeatedly been observed to be more sensitive to cannabinoid-induced antinociception compared to males.[10,11] As such, the female ECS is distinct with respect to the pharmacokinetic and pharmacodynamic properties allowing for greater cannabinoid-induced antinociceptive potency and/or efficacy. Notwithstanding increasing efforts to include female subjects in preclinical endocannabinoid research, there less consideration for how female endocrinology drives these sexual dimorphisms. Specifically, interactions between the ECS and fluctuations of female sex hormones—termed the estrus cycle in female rodents—is often overlooked in sex differences-focused studies.

Comparable to the 28-day human menstrual cycle, female mice and rats undergo a 4–5 day estrus cycle consisting of 4 main phases: proestrus, estrus, metestrus, and diestrus.[12,13] Proestrus is characterized by a dramatic increase in circulating estrogens or estradiol (E2) followed by a spike in the other hormones, namely progesterone, luteinizing hormone (LH), and follicle stimulating hormone (FSH).[12,13] In the beginning of estrus, E2 maintains moderate-to-high levels as compared to progesterone, LH, and FSH, which are at their lowest at the beginning of estrus.[12,13] The last 2 phases, metestrus and diestrus, are marked by decreased circulating E2, LH, and FSH while progesterone levels rise.[12,13] Both organizational (via mature sex organs/ gonads) and activational (brain-derived) forces drive these hormone fluctuations[14,15] as preclinical researchers often perform gonadectomies and/or hormone replacement treatments to control these concentrations in *in vivo* experiments. As with most bodily organs, the brain is largely affected by these hormone fluctuations. Particularly, E2 influences ECS functionality differently based on the brain region.[16–20] This results in unique profiles of cannabinoid-induced behaviors across the estrus cycle.[11,15]

To date, there are only a handful of published reports detailing how estus cycle phase affects cannabinoid receptors and endocannabinoids in different forebrain, midbrain, and hindbrain regions of mice and rats.[16–20] Some of these studies measure distinct aspects of cannabinoid receptor expression or functionality (e.g., mRNA transcription, protein quantification, binding site density, and binding affinity) while others focus on brain region-specific concentrations of 2-AG and AEA.[17,18,20] The earliest report by Rodríguez de Fonseca et al.[16] showed that in the hypothalamus of rats, non-discriminant CB1R/CB2R density and affinity are greatest during diestrus—when circulating E2 is low—as compared to estrus, when levels are elevated. As for hypothalamic endocannabinoids in rats, both 2-AG and AEA concentrations are also increased during diestrus as compared to the other cycle phases.[18] CB1R mRNA transcript levels in the pituitary are positively associated with circulating E2 in rats,[17] while pituitary endocannabinoid concentrations are less in sync with its' receptor patterns.[18] Data on the other brain regions remain scattered and/or inconsistent between publications despite their shared goal to make sense of these regional differences (Tables 1 and 2).[16–20]

The current review article aims to provide a comprehensive picture of how the ECS is modulated by the estrus cycle in different brain regions. A systematic approach will be used to compile and summarize all brain region-specific cannabinoid receptor, 2-AG, and AEA data.

Table 1 Brain region-specific cannabinoid receptor modulation by the rodent estrus cycle.

General region	Specific region	Reference	Species	Cannabinoid receptors			
				Proestrus	Estrus	Metestrus	Diestrus
Forebrain		Rodríguez de Fonseca et al.[16]	Rat	− CBR density, affinity	↓ CBR affinity	− CBR density, affinity	↑ CBR affinity
	Hypothalamus	Liu et al.[20]	Mouse	− CB1R in mRNA	− CB1R in mRNA		
	MBH	Rodríguez de Fonseca et al.[16]	Rat	− CBR density, affinity	↓ CBR density	− CBR density, affinity	↑ CBR density
	Anterior Pituitary	González et al.[17]		− CB1R mRNA	↓ CB1R mRNA	− CB1R mRNA	↑ CB1R mRNA
	Thalamus	Liu et al.[20]	Mouse	− CB1R mRNA	− CB1R mRNA		
	Hippocampus						
	Striatum	Rodríguez de Fonseca et al.[16]	Rat	− CBR density, affinity	− CBR density, affinity		
	NAC	Liu et al.[20]	Mouse	− CB1R mRNA	− CB1R mRNA		
	Medial Amygdala			↓ CB1R mRNA	↑ CB1R mRNA	↓ CB1R mRNA	
	BLA						
	BNST			− CB1R mRNA			
	Cerebral Cortex						

		Rodríguez de Fonseca et al. [16] Rat	Liu et al. [20] Mouse
Midbrain	Colliculus	– CBR density, affinity	– CB1R mRNA
	PAG		
	PBN	↓ CB1R mRNA	↑ CB1R mRNA
	VTA		↓ CB1R mRNA
Hindbrain	PPRF	– CB1R mRNA	
	RTTG		
	LC		

Arrows ↑ and ↓ represent statistically significant increases and decreases, respectively. The symbol – corresponds to no statistically significant change relative to the other estrus cycle phases. CBR, cannabinoid receptor; CB1R, cannabinoid receptor type 1; MBH, medial basal hypothalamus; NAC, nucleus accumbens; BLA, basolateral amygdala; BNST, bed nucleus of the stria terminalis; PAG, periaqueductal gray; PBN, parabrachial nucleus; VTA, ventral tegmental area; PPRF, paramedian pontine reticular formation; RTTG, reticulotegmental nucleus of the pons; LC, locus coeruleus.

Data was compiled and summarized from Rodríguez de Fonseca F, Cebeira M, Ramos JA, Martín M, Fernández-Ruiz JJ. Cannabinoid receptors in rat brain areas: sexual differences, fluctuations during estrous cycle and changes after gonadectomy and sex steroid replacement. *Life Sci.* 1994;54:159–170; González S, et al. Sex steroid influence on cannabinoid CB(1) receptor mRNA and endocannabinoid levels in the anterior pituitary gland. *Biochem Biophys Res Commun.* 2000;270:260–266; Levine A, et al. Sex differences in the expression of the endocannabinoid system within V1M cortex and PAG of sprague dawley rats. *Biol Sex Differ.* 2021;12:60.

Table 2 Brain region-specific endocannabinoid modulation by the rat estrus cycle.

General region	Specific region	Reference	Species	Endocannabinoids			
				Proestrus	Estrus	Metestrus	Diestrus
Forebrain	Hypothalamus	Bradshaw et al.[18]	Rat	↓ 2-AG	– 2-AG	↓ 2-AG	↑ 2-AG
				↑ AEA			↑ AEA
		González et al.[17]		– AEA			
	Pituitary	Bradshaw et al.[18]		↑ 2-AG	↓ 2-AG	– 2-AG	
				↑ AEA	– AEA	↓ AEA	– AEA
		González et al.[17]		↓ AEA	↑ AEA	↓ AEA	
	Thalamus	Bradshaw et al.[18]		– 2-AG, AEA			
	Hippocampus			– 2-AG			
				↓ AEA	↑ AEA	– AEA	
	Striatum			– 2-AG, AEA			
	V1	Levine et al.[20]		↓ 2-AG:AEA	↑ 2-AG:AEA	↓ 2-AG:AEA	↑ 2-AG:AEA
				↑ AEA	↓ AEA	↑ AEA	
Midbrain		Bradshaw et al.[18]		↑ 2-AG	↓ 2-AG		
				↑ AEA	↓ AEA		– AEA
	PAG	Levine et al.[20]		– 2-AG			
				– AEA		↑ AEA	– AEA
Hindbrain	Cerebellum	Bradshaw et al.[18]		– 2-AG nor AEA			
	TNC	Levine et al.[20]		↓ 2-AG		– 2-AG	↑ 2-AG
				↓ 2-AG:AEA	– 2-AG:AEA		↑ 2-AG:AEA
				– AEA			

Arrows ↑ and ↓ represent statistically significant increases and decreases, respectively. The symbol – corresponds to no statistically significant change relative to the other estrus cycle phases. 2-AG, 2-arachidonoylglycerol; AEA, anandamide; V1, primary visual cortex; PAG, periaqueductal gray; TNC, trigeminal nucleus caudalis.

...in cannabinoid CB1 receptor mRNA and endocannabinoid levels in the anterior pituitary gland. *Biochem Biophys Res Commun.* 2000;270:260–266; Bradshaw

The presentation of this data in tables will (1) enable readers to pinpoint information from specific regions of the forebrain, midbrain, and hindbrain; and (2) allow visualization of regional trends to inspire future research questions. The second part of this review article will focus on the molecular and cellular processes by which E2 affect the ECS. The last decade has seen critical advancements in delineating these mechanisms in the hypothalamus, hippocampus, and amygdala of rodents.[21-25] In short, E2 activates either estrogen receptor alpha (ERα) and beta (ERβ) heterodimers on the plasma membrane leading to postsynaptic endocannabinoid release and retrograde signaling at presynaptic γ-aminobutyric acid (GABA)-ergic neurons in some sex-specific brain regions, but not all.[21-25] Thus, relative plasma membrane ER distribution and synaptic concentrations of E2 determine specific thresholds of ER activity indicative of subsequent ECS functionality.

2. Literature search methodology

2.1 Data source and search strategy

In order to compose Tables 1 and 2, articles were sourced from PubMed from the time of the current review article's inception, February 18, 2022, to March 31, 2022. The following groups of terms were searched on PubMed to find studies to make Table 1: (estrus cycle) AND (cannabinoid receptor) which yielded 10 results from 1993 to 2021, and (estrous cycle) AND (cannabinoid receptor), which produced 26 results from 1992 to 2022. The following groups of terms were searched on PubMed to identify studies and datasets for the composition of Table 2 (estrus cycle) AND (endocannabinoid) which yielded the 9 results from 2000 to 2022; and (estrous cycle) AND (endocannabinoid), which produced 20 results from 2006 to 2022. The screening of results was undertaken by 2 independent reviewers (HJK and SLB), who systematically assessed each article and its dataset based on the inclusion criteria, first by title and abstract, then by full text including methods and results. Publication date was not restricted, while both searchers were constrained to English-language articles only. HJK and SLB compared their lists of which articles and datasets should be used for Tables 1 and 2. Any disagreements were resolved by discussion-based consensus through a third independent reviewer (TB). The searches were updated by HJK, SLB, and TB on March 31, 2022. This review was conducted in accordance with the Preferred Reported Items for Systematic Reviews and Meta-analyses (PRISMA) guidelines.[26]

The studies and datasets included in the Tables 1 and 2 were evaluated based on the following inclusion criteria: The included articles were of original research; therefore, review articles were not included. The animal species from which the data was derived were either mice or rats. All evidence was from sexually mature, normally cycling females and not males. Furthermore, these females were not experimentally modified prior to data collection. The type of bodily samples used for data collection and analysis was brain tissue. The following datasets were considered from female brain tissue: CB1R/CB2R mRNA transcription/expression, CB1R/CB2R binding site density, CB1R/CB2R protein expression, CB1R/CB2R binding affinity to cannabinoid receptor ligands (i.e., CP55,940, THC), and 2-AG/AEA concentrations. Following table composition, information for the remaining sections of this review article were sourced from PubMed and Google Scholar.

3. Brain region-specific ECS modulation by the estrus cycle

3.1 Components of ECS functionality

Cannabinoid receptors are the most ubiquitous GPCRs in the CNS. At the individual cell level, activation of these $G\alpha_{i/o}$ protein-coupled receptors leads to intracellular K^+ efflux, the closure of voltage-gated Ca^{2+} channels (VGCC), neuronal hyperpolarization, and the inhibition of action potentials [reviewed in[27,28]]. However, based on the precise location of CB1R and CB2R (i.e., expression on presynaptic/postsynaptic plasma membrane, GABAergic/glutamatergic neurons, inputs/outputs to different brain regions) the ECS serves important fine-tuning roles within larger inhibitory and excitatory circuits. For example, endocannabinoids may either bind receptors located on postsynaptic or presynaptic membranes. In the latter scenario, termed retrograde signaling, postsynaptically synthesized 2-AG and AEA transverse the synaptic cleft to bind presynaptic CB1R on GABAergic neurons.[27,28] As already mentioned, CB1R activation results in hyperpolarization and inhibition of GABA release.[27,28] This leads to decreased synaptic concentrations of GABA, increased glutamatergic binding at postsynaptic receptors, and overall disinhibition or excitatory transmission.[27,28] These intra- and extracellular events are depicted in Fig. 3. Neuromodulatory systems such as the estrogen signaling system influences CB1R-mediated retrograde signaling by promoting or inhibiting postsynaptic endocannabinoid synthesis.[21–23] These molecular mechanisms give clues to how the estrus cycle and its associated E2 fluctuations determine sex- and brain-region specific ECS functionality.

Brain region-specific ECS functionality also depends on relative cannabinoid receptor subtype distribution as well as pre-existing synaptic abundance of 2-AG and AEA. Both cannabinoid receptor subtypes are largely expressed throughout the CNS and periphery. For example, CB1R is especially dense in the cerebral cortex, hippocampus, basal ganglia, cerebellum, and brain stem[29–31] despite also having function in hepatocytes[32] and adipocytes.[33,34] Despite CB2R being abundantly expressed throughout the periphery,[35] this receptor subtype also operates on glial cells such as astrocytes and microglia.[36,37] When it comes to the endocannabinoids, 2-AG exists in 170-times greater concentrations than AEA.[38] 2-AG acts as a full agonist at both cannabinoid receptor subtypes.[39] Although AEA exists in much smaller concentrations in the brain, it has a high affinity for CB1R and a lower affinity for CB2R.[39] As such, AEA acts as a partial-to-full agonist at CB1R while it is nearly inactive at CB2R.[39] These brain region-specific distribution and endocannabinoid binding details are important for interpreting estrus cycle-dependent patterns.

3.2 Brain region-specific patterns in endocannabinoid functionality

3.2.1 Forebrain

In the greater limbic forebrain of rats, CB1R/CB2R binding affinity is elevated during diestrus as compared to estrus,[16] whereas endocannabinoid concentrations vary across specific forebrain regions including the hypothalamus, pituitary, thalamus, hippocampus, striatum, and primary visual cortex (V1). In the medial basal hypothalamus (MBH) of rats, CB1R/CB2R density is greatest in diestrus[16] along with concentrations of both endocannabinoids.[17,18] In the anterior pituitary, CB1R mRNA expression is also largest during rat diestrus. However, with regards to the pituitary's endocannabinoid concentrations, 2-AG levels are higher in proestrus as compared to estrus in rats,[18] while AEA concentrations are the greatest during estrus.[17] CB1R expression does not change in the thalamus and hippocampus throughout estrus cycle phases in mice.[19] Regarding the endocannabinoids, rat hippocampal concentrations of AEA are greater in estrus as compared to proestrus.[18] Within the striatum and nucleus accumbens (NAC), the estrus cycle does not affect CB1R/CB2R density, affinity, nor CB1R mRNA transcription in rats.[16,19] Striatal levels of endocannabinoids are also unaffected by cycle phases.[18]

The medial and basolateral (BLA) regions of the amygdala show a different pattern to all other brain regions. Unlike the hypothalamus and pituitary, CB1R mRNA levels in the mouse amygdala are highest during estrus as compared to any other cycle phase.[19] Unfortunately, the corresponding endocannabinoid data for the amygdala does not exist. As an extension of the amygdala, the bed nucleus of the stria terminalis (BNST) of mice does not undergo estrus cycle-dependent changes in CB1R mRNA expression.[19] Moreover, neither does the mouse cerebral cortex with respect to CB1R mRNA transcription.[19] Specifically within the V1 of rats, the concentration ratio between 2-AG and AEA (2-AG:AEA) is greatest in estrus as compared to the other cycle phases.[20] AEA levels are elevated during proestrus, diestrus, and metestrus, while they are lowest during estrus in the V1.[20] All of these forebrain-specific ECS patterns are shown in Tables 1 and 2, which correspond to the visual representation of the brain regions in Figs. 1 and 2.

3.2.2 Midbrain

There is less cannabinoid receptor and endocannabinoid data on midbrain regions. Throughout the midbrain, neither CB1R/CB2R density nor affinity are affected by the estrus cycle in rats.[16] Despite this, 2-AG concentrations are highest during proestrus relative to all other estrus cycle phases in rats.[18] Moreover, the greater midbrain's AEA concentrations are also

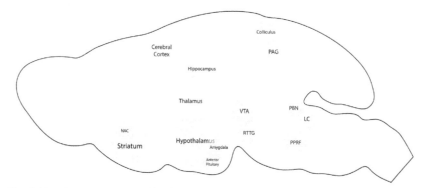

Fig. 1 Visual representation of rodent brain region-specific cannabinoid receptor modulation by the estrus cycle. The regions depicted are described in detail in Table 2 for CB1R-specific studies and studies where CB1R and CB2R are not differentiated. NAC, nucleus accumbens; PAG, periaqueductal gray; PBN, parabrachial nucleus; VTA, ventral tegmental area; PRRF, paramedian pontine reticular formation; RTTG, reticulotegmental nucleus of the pons; LC, locus coeruleus.

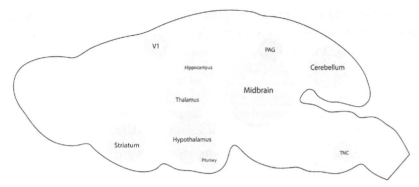

Fig. 2 Visual representation of rodent brain region-specific endocannabinoid modulation by the estrus cycle. The regions depicted are described in detail in Table 1 for both AEA and 2-AG. V1, primary visual cortex; PAG, periaqueductal gray; TNC, trigeminal nucleus caudalis.

elevated during proestrus as compared to both estrus and metestrus.[18] CB1R mRNA expression in the colliculus—a structure involved in transforming sensory information into movement output[40]—does not change across the estrus cycle in mice.[19] CB1R mRNA transcription within the periaqueductal gray (PAG), a key region for endocannabinoid-mediated antinociception,[41] is also unchanged throughout the mouse estrus cycle.[19] When it comes to AEA within the parabrachial nucleus (PBN) of rats, levels are greater in metestrus than in estrus.[20] Finally, cycle–dependent differences in CB1R mRNA have not been observed in the mouse ventral tegmental area (VTA), the hub of dopaminergic outputs to mesocortical and mesolimbic circuits[19,42] nor is there any information on the endocannabinoids in this region. Tables 1 and 2 contain all the described receptor and ligand data for the midbrain regions, respectively. Figs. 1 and 2 help visualize region-specific data.

3.2.3 Hindbrain
There is the least amount of data on this subject for hindbrain regions. There are no CB1R mRNA differences in the paramedian pontine reticular formation (PPRF), reticulotegmental nucleus of the pons (RTTG), nor locus coeruleus (LC)[19] across the mouse estrus cycle. Spanning from the pons to the mid-medulla, the trigeminal nucleus caudalis (TNC) has increased 2-AG during diestrus as compared to proestrus and estrus in rats.[20] Furthermore, the ratio of 2-AG to AEA is greater during diestrus than in proestrus.[20] There is no estrus cycle-dependent CB1R/CB2R information on the

cerebellum. Concerning endocannabinoid concentrations in the rat cerebellum, they do not change as a function of cycle phase in rats.[18] All paramedian PPRF, RTTG, LC, TNC, and cerebellar ECS evidence is organized in Tables 1 and 2 and illustrated in Figs. 1 and 2.

4. Mechanisms underlying brain region-specific endocannabinoid modulation by the estrus cycle

4.1 E2-modulated endocannabinoid synthesis in the hypothalamus

Valuable E2-related insights can be gained from comparing males, gonadectomized females, and sex hormone-treated animals. Within the hypothalamus of rats, CB1R/CB2R density is greater in females than males, while ovariectomized females that embody lower E2 levels display increased receptor binding compared to their ovary-intact counterparts.[43,44] Thus, hypothalamic AEA levels and CB1R affinity are negatively associated with E2 abundance.[16,18,44] Hypothalamic 2-AG of mice, on the other hand, does not exemplify this pattern.[23] E2 application to median preoptic nucleus slices of ovary-intact metestrus mice reduces the frequency of spontaneous postsynaptic currents.[23] E2 acts through ERβ located on the plasma membrane to activate phospholipase C (PLC) leading to the generation of 1,4,5-trisphosphate (IP3) from phosphatidylinositol bisphosphate (PIP2).[23] IP3 then stimulates the endoplasmic reticulum resulting in intracellular Ca^{2+} release and diacylglycerol (DAG) synthesis of 2-AG.[23] Fig. 3 illustrates this intracellular cascade. It is unclear whether ERβ functions singularly or as a heterodimer in these experiments,[23] as non-dimerized ERs typically precipitate the adenylyl cyclase (AC)/adenosine monophosphate (AMP)/cyclic adenosine monophosphate (cAMP) pathway leading to neuronal hyperpolarization.[45]

4.2 E2-modulated endocannabinoid synthesis in the hippocampus

In the rat hippocampus, 2-AG content correlates positively with E2 during estrus, while AEA concentrations remain unaffected.[18] Despite this, hippocampal recordings of GABA receptor (GABAR)-mediated inhibitory postsynaptic currents have shown that E2 influences AEA synthesis and synaptic abundance.[21,22] E2 binds heterodimerized ERα and metabotropic glutamate receptor type 1 (mGluR1; ERα-mGluR1) on the postsynaptic plasma membrane to also trigger the PLC/PIP2/IP3 pathway.[21,22]

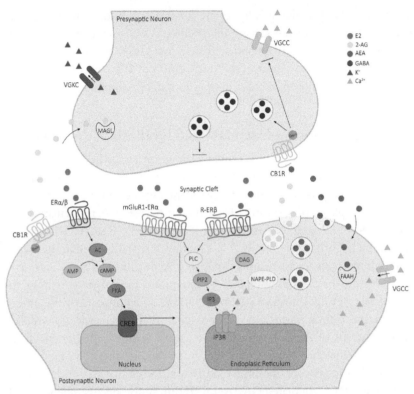

Fig. 3 E2 activation of non-dimerized and heterodimerized ERs affecting downstream endocannabinoid synthesis, synaptic release, retrograde signaling, and CB1R-mediated GABAergic disinhibition. E2, estradiol; ER, estrogen receptor; ERα, estrogen receptor alpha; ERβ, estrogen receptor beta; AC, adenylyl cyclase; AMP, adenosine monophosphate; cAMP, cyclic adenosine monophosphate; PKA, protein kinase A; CREB, cAMP response element binding protein; mGluR1, metabotropic glutamate receptor; PLC, phospholipase C; PIP2, phosphatidylinositol bisphosphate; IP3, 1,4,5-trisphosphate; VGCC, voltage-gated Ca^{2+} channel; IP3R, 1,4,5-trisphosphate receptor; DAG, diacylglycerol; NAPE-PLD, N-acyl phosphatidylethanolamine-specific phospholipase D; 2-AG, 2-arachidonoylglycerol; MAGL, monoacylglycerol lipase; AEA, anandamide; FAAH, fatty acid amide hydrolase; CB1R, cannabinoid receptor type 1; $G\alpha_{i/o}$, $G_{i/o}$ protein alpha subunit; VGKC, voltage-gated K^+ channel; GABA, γ-aminobutyric acid.

However, instead of DAG activation, N-acyl phosphatidyletholamine-specific phospholipase D (NAPE–PLD) is recruited to synthesize AEA.[22] These alternative intracellular events are represented in Fig. 3. AEA then activates presynaptically expressed CB1R via retrograde signaling leading to GABAergic disinhibition and subsequent glutamatergic excitation.[21] Remarkably, E2's regulation of ERα–mGluR1 heterodimers only occurs in females.[22] Fatty acid amide hydrolase (FAAH) inhibitor administration to female CA1 slices prevents more than 50% of CB1R-mediated inhibitory responses; however, this synaptic effect is not observed in male slices.[22] These results call into question the distribution of ERα-mGluR1 as compared to the non-dimerized ERs in sex- and brain region-specific synapses.

4.3 E2-modulated endocannabinoid synthesis in the amygdala

Given E2's modulation of endocannabinoid synthesis leading to presynaptic CB1R activation, it is reasonable to assume that the estrus cycle and its associated sex hormone fluctuations has some impact on CB1R reorganization. This explains why during proestrus and estrus—when circulating and brain levels of E2 are high—there is downregulated CB1R/CB2R expression in the hypothalamus of rats.[16,17] The opposite is observed in the mouse amygdala of both sexes, relating back to ER distribution and functionality.[19,24,25] In the central amygdala of estrus and diestrus rats, neither the CB1R agonist WIN-55,212-2 nor CB1R antagonist AM251 affects GABAergic disinhibition as to alter excitatory postsynaptic potentials.[24] In the nucleus of the stria terminalis—a tract that carries fibers from the amygdala to the septal nuclei, hypothalamus, and thalamus—E2 rapidly elevates inhibitory synaptic transmission in both sexes of mice.[25] Although it is likely that CB1R-mediated GABAergic disinhibition occurs independently of E2 signaling in the central amygdala,[24] its output fiber may be subject to receptor-mediated interference of endocannabinoid synthesis.[25]

In contrast to hypothalamic and hippocampal synapses, amygdalar plasma membranes may contain higher concentrations of ERs that evoke contrasting transmission patterns as compared to the other brain regions.[46–48] Increased activation of non-dimerized ERα and/or ERβ stimulates adenylyl cyclase (AC) to convert adenosine monophosphate (AMP) into cyclic adenosine monophosphate (cAMP).[45] Next, protein kinase A (PKA) is stimulated to bind cAMP response element binding protein (CREB) on the nuclear membrane.[45] This presumably decreases DAG and/or NAPE-PLD synthesis of endocannabinoids, packaging in vesicles, and release into the synaptic cleft.[45] Fig. 3 summarizes the scenario wherein E2 signaling interferes with endocannabinoid synthesis to prevent CB1R-mediated GABAergic disinhibition in the nucleus of the stria terminalis and to a lesser degree, in the central amygdala. Unfortunately, there is no estrus cycle-dependent endocannabinoid data on either the amygdala or nucleus of the stria terminalis.

4.4 Estrogen fluctuations

Similar to the ECS, estrogen system functionality is also brain region-specific. E2 concentrations in the hippocampus of female rats in proestrus, estrus, diestrus-1, and diestrus-2/metestrus are 1.7 nM, 1.0 nM, 0.5 nM, and 0.7 nM, respectively.[49] Conversely, the male hippocampus contains approximately 8.4 nM or 5–12 times more E2 compared to the hippocampus of females, a large portion of which is locally produced from androstenedione metabolism.[49] Within the female rat brain, ERβ mRNA is more highly expressed in the cerebral cortex and hippocampus as compared to ERα.[50] As for the hypothalamus, the lateral region is exclusive in ERβ-positive neurons while the ventromedial hypothalamus strictly expresses ERα mRNA.[50] And although there are no studies comparing ER-mGluR1 heterodimer distribution between female brain regions, female rat brain expression of mGluR1 mRNA is largest in the basal ganglia, preoptic nucleus of the hypothalamus, and dorsal cochlear nucleus of the hindbrain.[51] It may be that higher expression and function of heterodimerized ERβ and ERα-mGluR1 correlates to greater CB1R-mediated GABAergic disinhibition. In the case where receptor distribution is unchanged between brain regions, the varying degrees of endocannabinoid retrograde signaling may be due to differential binding profiles (i.e., allosterism, affinity) between non-dimerized and hetero-dimerized ERs. Coming back to the electrophysiological studies by Gregory and colleagues,[25] E2 was 100-fold more potent in inhibiting synaptic transmission in female stria terminalis slices relative to that in male mice. However, GABA-mediated inhibitory transmission was not significantly different between female mice in estrus compared to diestrus in the presence of 1 nM E2.[25] These results indicate that synaptic E2 concentrations between 1.5–2.7 nM [(0.5 nM during diestrus + 1 nM) − (1.7 nM during proestrus + 1 nM)] do not directly evoke changes in inhibitory transmission across the phases of the estrus cycle,[25] implying that E2 flux is not responsible for changes in inhibitory neurotransmission, but may facilitate changes through other systems, such as the ECS.

5. Limitations and future directions

There are a number of gaps in the literature preventing us from developing a complete understanding of the relationship between the ECS, E2, and the estrus cycle. First, E2 undergoes metabolism in the synaptic cleft, as it is broken down into estriol and estrone.[52] E2 is 80- and 12-times more potent than estriol and estrone, respectively,[53] which explains why higher synaptic concentrations of E2—and subsequent breakdown into these metabolites—results in decreased synaptic inhibition.[25] It remains unknown whether E2, estriol, and estrone have differing affinities or binding patterns at non-dimerized and heterodimerized ERs. There are also no published investigations on 2-AG degrading enzyme monoacylglycerol acid lipase, while there is a single study on FAAH which shows that E2's anxiolytic effects are contingent on CB1R functionality in rats.[54] Further related to receptor and ligand abundances, the metrics of receptor reorganization, including (1) the levels of activation required to initiate these events, (2) rate, and (3) overall efficiency of receptor downregulation or upregulation *in vivo* are not well defined. These details would improve understanding between the cannabinoid receptor and endocannabinoid data presented in Tables 1 and 2. Finally, our literature search yielded no data on the role of CB2R. Future studies should not only focus on CB2R, but also apply the current molecular knowledge to pain processing circuits. The rodent estrus cycle affects anti-nociception more so than other cannabinoid-induced behaviors.[11,15,55,56] Better understanding of how ER heterodimer-mediated endocannabinoid retrograde signaling occurs in the context of pain is therefore the next logical step towards translatability.

References

1. Fernández-Ruiz J. The endocannabinoid system as a target for the treatment of motor dysfunction. *Br J Pharmacol.* 2009;156:1029–1040.
2. Zanettini C, Panlilio LV, Alicki M, Goldberg SR, Haller J, Yasar S. Effects of endocannabinoid system modulation on cognitive and emotional behavior. *Front Behav Neurosci.* 2011;5:57.
3. Solinas M, Goldberg SR, Piomelli D. The endocannabinoid system in brain reward processes. *Br J Pharmacol.* 2008;154:369–383.
4. Woodhams SG, Sagar DR, Burston JJ, Chapman V. The role of the endocannabinoid system in pain. *Handb Exp Pharmacol.* 2015;227:119–143.
5. Morena M, Patel S, Bains JS, Hill MN. Neurobiological interactions between stress and the endocannabinoid system. *NPP.* 2016;41:80–102.
6. Kesner AJ, Lovinger DM. Cannabinoids endocannabinoids and sleep. *Front Mol Neurosci.* 2020;13:125.
7. Rahman S, Uyama T, Hussain Z, Ueda N. Roles of endocannabinoids and endocannabinoid-like molecules in energy homeostasis and metabolic regulation: a nutritional perspective. *Annu Rev Nutr.* 2021;41:177–202.
8. Mackie K. Cannabinoid receptors: where they are and what they do. *J Neuroendocrinol.* 2008;20:10–14.
9. Zou S, Kumar U. Cannabinoid receptors and the endocannabinoid system: signaling and function in the central nervous system. *Int J Mol Sci.* 2018;19:833.

10. Tseng AH, Craft RM. Sex differences in antinociceptive and motoric effects of cannabinoids. *Eur J Pharmacol*. 2001;430:41–47.

11. Wakley AA, Craft RM. Antinociception and sedation following intracerebroventricular administration of Δ^9-tetrahydrocannabinol in female vs. male rats. *Behav Brain Res*. 2011;216:200–206.

12. Butcher RL, Collins WE, Fugo NW. Plasma concentration of LH, FSH, prolactin, progesterone and estradiol-17beta throughout the 4-day estrous cycle of the rat. *Endocrinology*. 1974;94:1704–1708.

13. Miller BH, Takahashi JS. Central circadian control of female reproductive function. *Front Endocrinol*. 2014;4:195.

14. Becker JB, et al. Strategies and methods for research on sex differences in brain and behavior. *Endocrinology*. 2005;146:1650–1673.

15. Craft RM, Leitl MD. Gonadal hormone modulation of the behavioral effects of Delta9-tetrahydrocannabinol in male and female rats. *Eur J Pharmacol*. 2008;578:37–42.

16. Rodríguez de Fonseca F, Cebeira M, Ramos JA, Martín M, Fernández-Ruiz JJ. Cannabinoid receptors in rat brain areas: sexual differences, fluctuations during estrous cycle and changes after gonadectomy and sex steroid replacement. *Life Sci*. 1994;54:159–170.

17. González S, et al. Sex steroid influence on cannabinoid CB(1) receptor mRNA and endocannabinoid levels in the anterior pituitary gland. *Biochem Biophys Res Commun*. 2000;270:260–266.

18. Bradshaw HB, Rimmerman N, Krey JF, Walker JM. Sex and hormonal cycle differences in rat brain levels of pain-related cannabimimetic lipid mediators. *Am J Physiol Regul Integr Comp Physiol*. 2006;291:349–358.

19. Liu X, Li X, Zhao G, Wang F, Wang L. Sexual dimorphic distribution of cannabinoid 1 receptor mRNA in adult C57BL/6J mice. *J Comp Neurol*. 2020;528:1986–1999.

20. Levine A, et al. Sex differences in the expression of the endocannabinoid system within V1M cortex and PAG of Sprague Dawley rats. *Biol Sex Differ*. 2021;12:60.

21. Huang GZ, Woolley CS. Estradiol acutely suppresses inhibition in the hippocampus through a sex-specific endocannabinoid and mGluR-dependent mechanism. *Neuron*. 2012;74:801–808.

22. Tabatadze N, Huang G, May RM, Jain A, Woolley CS. Sex differences in molecular signaling at inhibitory synapses in the hippocampus. *J Neurosci*. 2015;35:11252–11265.

23. Bálint F, Liposits Z, Farkas I. Estrogen receptor beta and 2-arachidonoylglycerol mediate the suppressive effects of estradiol on frequency of postsynaptic currents in gonadotropin-releasing hormone neurons of metestrous mice: an acute slice electrophysiological study. *Front Cell Neurosci*. 2016;10:77.

24. Kirson D, Oleata CS, Parsons LH, Ciccocioppo R, Roberto M. CB1 and ethanol effects on glutamatergic transmission in the central amygdala of male and female msP and wistar rats. *Addict Biol*. 2018;23:676–688.

25. Gregory JG, Hawken ER, Angelis S, Bouchard JF, Dumont ÉC. Estradiol potentiates inhibitory synaptic transmission in the oval bed nucleus of the striaterminalis of male and female rats. *Psychoneuroendocrinology*. 2019;106:102–110.

26. Moher D, Liberati A, Tetzlaff J, Altman DG, PRISMA Group. Preferred reporting items for systematic reviews and meta-analyses: the PRISMA statement. *PLoS Med*. 2009;6:1000097.

27. Vaughan CW, Christie MJ. Retrograde signalling by endocannabinoids. *Handb Exp Pharmacol*. 2005;168:367–383.

28. Howlett AC, Blume LC, Dalton GD. CB(1) cannabinoid receptors and their associated proteins. *Curr Med Chem*. 2010;17:1382–1393.

29. Herkenham M, et al. Characterization and localization of cannabinoid receptors in rat brain: a quantitative in vitro autoradiographic study. *J Neurosci*. 1991;11:563–583.

30. Mackie K. Distribution of cannabinoid receptors in the central and peripheral nervous system. *Handb Exp Pharmacol.* 2005;168:299–325.

31. Marsicano G, Kuner R. Anatomical distribution of receptors, ligands and enzymes in the brain and in the spinal cord: circuitries and neurochemistry. In: Köfalvi A, ed. *Cannabinoids and The Brain.* Boston, MA: Springer; 2008:161–201.

32. Chanda D, et al. Cannabinoid receptor type 1 (CB1R) signaling regulates hepatic gluconeogenesis via induction of endoplasmic reticulum-bound transcription factor cAMP-responsive element-binding protein H (CREBH) in primary hepatocytes. *Int J Biol Chem.* 2011;286:27971–27979.

33. Behl T, et al. Understanding the possible role of endocannabinoid system in obesity. *Prostaglandins Other Lipid Mediat.* 2021;152:106520.

34. Gasperi V, et al. Endocannabinoids in adipocytes during differentiation and their role in glucose uptake. *Cell Mol Life Sci.* 2007;64:219–229.

35. Onaivi ES, et al. Expression of cannabinoid receptors and their gene transcripts in human blood cells. *Prog Neuropsychopharmacol Biol Psychiatry.* 1999;23:1063–1077.

36. Cabral GA, Raborn ES, Griffin L, Dennis J, Marciano-Cabra F. CB2 receptors in the brain: role in central immune function. *Br J Pharmacol.* 2008;153:240–251.

37. Atwood BK, Mackie K. CB2: a cannabinoid receptor with an identity crisis. *Br J Pharmacol.* 2010;160:467–479.

38. Stella N, Schweitzer P, Piomelli D. A second endogenous cannabinoid that modulates long-term potentiation. *Nature.* 1997;388:773–778.

39. Di Marzo V, De Petrocellis L. Why do cannabinoid receptors have more than one endogenous ligand? *Philos Trans R Soc.* 2012;367:3216–3228.

40. Wurtz RH. Superior colliculus. In: Squire LR, ed. *Encyclopedia of Neuroscience;* 2009:627–634.

41. Lichtman AH, Cook SA, Martin BR. Investigation of brain sites mediating cannabinoid-induced antinociception in rats: evidence supporting periaqueductal gray involvement. *J Pharmacol Exp Ther.* 1996;276:585–593.

42. Merrill CB, Friend LN, Newton ST, Hopkins ZH, Edwards JG. Ventral tegmental area dopamine and GABA neurons: physiological properties and expression of mRNA for endocannabinoid biosynthetic elements. *Sci Rep.* 2015;5:16176.

43. Wise PM, Ratner A. Effect of ovariectomy on plasma LH, FSH, estradiol, and proges-terone and medial basal hypothalamic LHRH concentrations old and young rats. *Neuroendocrinology.* 1980;30:15–19.

44. Riebe CJ, Hill MN, Lee TT, Hillard CJ, Gorzalka BB. Estrogenic regulation of limbic cannabinoid receptor binding. *Psychoneuroendocrinology.* 2010;35:1265–1269.

45. Dobovišek L, Hojnik M, Ferk P. Overlapping molecular pathways between cannabinoid receptors type 1 and 2 and estrogens/androgens on the periphery and their involvement in the pathogenesis of common diseases (review). *Int J Mol Med.* 2016;38:1642–1651.

46. Lavreysen H, Pereira SN, Leysen JE, Langlois X, Lesage AS. Metabotropic glutamate 1 receptor distribution and occupancy in the rat brain: a quantitative autoradiographic study using [3H]R214127. *Neuropharmacology.* 2005;46:609–619.

47. Azad SC, et al. Circuitry for associative plasticity in the amygdala involves endo-cannabinoid signaling. *J Neurosci.* 2004;24:9953–9961.

48. Chen A, Hu WW, Jiang XL, Potegal M, Li H. Molecular mechanisms of group I meta-botropic glutamate receptor mediated LTP and LTD in basolateral amygdala in vitro. *Psychopharmacology.* 2017;234:681–694.

49. Hojo Y, et al. Hippocampal synthesis of sex steroids and corticosteroids: essential for modulation of synaptic plasticity. *Front Endocrinol.* 2011;2:43.

50. Shughrue PJ, Lane MV, Merchenthaler I. Comparative distribution of estrogen receptor-alpha and -beta mRNA in the rat central nervous system. *J Comp Neurol.* 1997;388:507–525.

51. Shigemoto R, Nakanishi S, Mizuno N. Distribution of the mRNA for a metabotropic glutamate receptor (mGluR1) in the central nervous system: an in situ hybridization study in adult and developing rat. *J Comp Neurol.* 1992;322:121–135.
52. Raftogianis R, Creveling C, Weinshilboum R, Weisz J. Estrogen metabolism by conjugation. *J Natl Cancer Inst.* 2000;27:113–124.
53. Hall JE, Hall ME. *Guyton and Hall Textbook of Medical Physiology e-Book.* Philadelphia, PA: Elsevier Health Sciences; 2015.
54. Hill MN, Karacabeyli ES, Gorzalka BB. Estrogen recruits the endocannabinoid system to modulate emotionality. *Psychoneuroendocrinology.* 2007;32:350–357.
55. Wakley AA, Wiley JL, Craft RM. Sex differences in antinociceptive tolerance to delta-9-tetrahydrocannabinol in the rat. *Drug Alcohol Depend.* 2014;143:22–28.
56. Kim HJJ, Zagzoog A, Black T, Baccetto SL, Ezeaka UC, Laprairie RB. Impact of the mouse estrus cycle on cannabinoid receptor agonist-induced molecular and behavioural outcomes. *Pharmacol Res Persp.* 2022. e-pub ahead of print.

CHAPTER THREE

Probing the orphan receptors: Tools and directions

Luca Franchini and Cesare Orlandi*

Department of Pharmacology and Physiology, University of Rochester Medical Center, Rochester, NY, United States
*Corresponding author: e-mail address: cesare_orlandi@urmc.rochester.edu

Contents

Abstract

The endogenous ligands activating a large fraction of the G Protein Coupled Receptor (GPCR) family members have yet to be identified. These receptors are commonly labeled as orphans (oGPCRs), and because of the absence of available pharmacological tools they are currently understudied. Nonetheless, genome wide association studies, together with research using animal models identified many physiological functions regulated by oGPCRs. Similarly, mutations in some oGPCRs have been associated with rare genetic disorders or with an increased risk of developing pathologies. The once underestimated pharmacological potential of targeting oGPCRs is increasingly being exploited by the development of novel tools to understand their biology and by drug discovery endeavors aimed at identifying new modulators of their activity. Here, we summarize recent advancements in the field of oGPCRs and future directions.

1. Introduction

G-protein coupled receptors (GPCRs) are 7-transmembrane domain proteins that sense extracellular events and transduce this information through a conformational change leading to activation of intracellular signaling cascades. Many GPCR family members are evolutionary conserved and mediate the effects of a plethora of extracellular stimuli such as odorants, light, neurotransmitters, and bioactive peptides. Being localized at the plasma membrane and involved in nearly every physiological process, GPCRs represent the gene family most targeted by available drugs.[1] However, a group of ~100 less-studied GPCRs, adding up to approximately a quarter of non-olfactory GPCRs, are known as "orphans" (oGPCRs) because their endogenous ligands have not yet been identified. Similarly, G protein coupling profile and signaling cascades initiated by many oGPCRs are not characterized. Nonetheless, disease-causing mutations identified in patients as well as studies on a variety of animal models have shown that many oGPCRs play crucial physiological roles, and thus, represent attractive drug targets.[2–6] Consequently, oGPCRs represent an important portion of the so-called dark druggable genome, a large group of understudied genes for which biochemical tools and assays are underdeveloped.[7] Recent efforts to systematically fill the gap in this research area led to the development of programs such as the International

Mouse Phenotype Consortium (IMPC), Illuminating the Druggable Genome consortium (IDG), Library of Integrated Network-Based Cellular Signatures consortium (LINCS), and the Structural Genomics Consortium (SGC).[7–10] According to the recommendations of the International Union of Basic and Clinical Pharmacology Committee on Receptor Nomenclature and Drug Classification (NC-IUPHAR), an oGPCR is considered de-orphanized when the following criteria are met[11,12]:

1. Reproducibility. At least two reports from independent research groups should provide indications to assign an endogenous ligand to an oGPCR. Ideally, proof of physical interaction together with evidence of receptor activation from functional assays should be provided;

2. *In vivo* pairing likelihood. Data should demonstrate the activation of the oGPCR by putative endogenous ligand with a potency that is consistent with its tissue concentration. Scientists are therefore recommended to provide a description of the mechanisms allowing the ligand to reach concentration levels compatible with receptor activation *in vivo*. The action of the candidate ligand is sometimes tested in tissues where the oGPCR expression has been ablated or increased. A significant difference in the measured signaling outcome supports the ligand-receptor pairing. However, long-term adaptations that affect expression and/or function of a range of signaling molecules after ablation of specific oGPCRs have been reported.[5,13,14] Such alterations may affect the observed readout resulting in inconclusive data interpretation. Similarly, formation of heteromers with oGPCRs have been shown to alter the signaling properties of canonical GPCRs.[15] Accordingly, it is essential to generate *in vitro* assays in heterologous expression systems that allow control over oGPCR membrane trafficking and accurate analysis of dose-response effects. Furthermore, ligand-independent functions are increasingly being reported and may provide important insights into the mechanisms of action of oGPCRs.[16,17] In this chapter, we focus on the most recently developed approaches to investigate oGPCRs (summarized in Fig. 1).

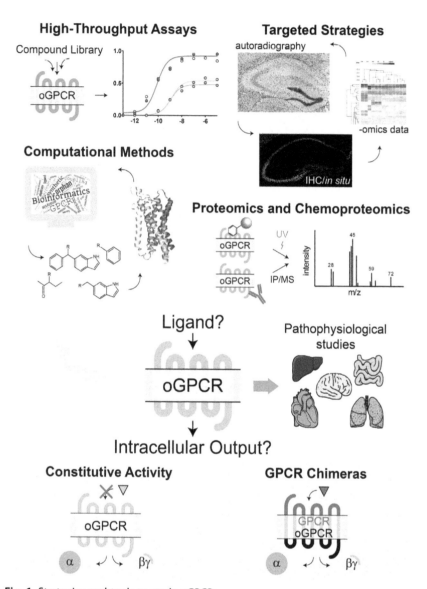

Fig. 1 Strategies and tools to study oGPCRs.

2. Defining the G protein coupling profile of oGPCRs

The quest for endogenous and synthetic ligands modulating the activity of oGPCRs is hampered by the lack of data about G protein coupling profiles and intracellular signaling cascades initiated by oGPCRs. Recent efforts directed to fill this gap of knowledge took advantage of sensitive *in vitro* assays to measure the constitutive activity of oGPCRs. At the plasma membrane, GPCRs oscillate between active and inactive structural conformations, with agonists stabilizing their active state to induce GTP exchange on $G\alpha$ subunits triggering the dissociation from $G\beta\gamma$ subunits and activation of intracellular effectors.[18] When a GPCR produces spontaneous G protein activation in the absence of agonist, constitutive activity can be measured, a phenomenon often observed in heterologous systems, but that can also be detected *in vivo*.[19–22] With the inactive state being the most stable and therefore more frequently represented, the ability to capture intracellular activity could be a daunting task requiring highly sensitive methods. To this aim, luciferase reporter assays have been frequently employed because of the high sensitivity provided by the intracellular accumulation of luciferase, especially after the recent development of brightest enzymes such as NanoLuc.[23–28] In a luciferase reporter assay, mammalian cells are co-transfected with the target GPCR and a plasmid encoding the luciferase enzyme downstream of a promoter activated by a specific G protein signaling cascade. Examples of such promoters are the Cyclic-AMP Responsive Element (CRE) to detect G_s activation, or the Nuclear Factor for Activation of T cells (NFAT) for G_q signaling. Activation of $G_{12/13}$ can be detected using promoters comprising a serum response element (SRE) or a serum response factor response element (SRF-RE), with the second one built to respond to SRF-dependent and ternary complex factor (TCF)-independent pathways.[29] A systematic use of this approach on a library of 41 GPCRs, including 19 orphans, successfully led to the detection of G_s, G_q, and $G_{12/13}$ constitutive activity for several control GPCRs.[30] However, according to data from the GPCRdb, more than 60% of the GPCRs signal through $G_{i/o/z}$ proteins, with half of them activating this class of G proteins uniquely,[31,32] and currently available assays are underdeveloped to measure intrinsically small constitutive $G_{i/o/z}$ activation. To overcome this issue, the authors used a combination of luciferase reporter assays and $G\alpha$ protein chimeras bearing the C-terminus of $G_{i/o/z}$ proteins. This approach was successful in redirecting G protein activation from $G_{i/o/z}$-coupled receptors to a readily measurable outcome such as CRE-dependent NanoLuc accumulation in cells.[30] In detail, $G\alpha$ protein chimeras are generated by swapping the last few amino acids (3 to 13) with those of a different $G\alpha$ protein.[21,33–35] This modification is sufficient to obtain a substantial activation of the $G\alpha$

chimera by receptors that would normally activate the G protein family corresponding to the C-terminus. For example, a GsGo chimera (core of $G\alpha_s$ and C-terminus of $G\alpha_o$) can be effectively activated by $G_{i/o/z}$-coupled μ-opioid receptor leading to easily quantifiable accumulation of intracellular cAMP.[34] Applying this strategy, the authors observed a significant $G_{i/o/z}$ coupling for 8 oGPCRs.[30] Alternative approaches employing newly developed Bioluminescence Resonance Energy Transfer (BRET) assays have been recently established to reliably detect constitutive activity in transfected cells.[36,37] Among the advantages of these strategies there is the direct measurement of G protein activation at the receptor bypassing the quantification of second messengers or transcriptional events. In fact, crosstalk between signaling pathways can be confounding and available assays are not always comparable across the four G protein families. In one of these studies oGPCR constitutive activity was measured with a BRET assay based on the GDP-disruption of the complex between oGPCRs and heterotrimeric G proteins.[36] In detail, it was recently shown that GPCRs can spontaneously couple to G proteins in absence of guanine nucleotides, while introducing GDP in the system was found sufficient to disrupt these interactions.[38] The BRET assay designed by Lu and colleagues had Rluc8 fused to the oGPCR C-terminus and the fluorescent protein Venus fused with Gβγ. Addition of 100 μM GDP to permeabilized cells co-transfected with unique Gα subunits belonging to the four main families (G_s, G_{i1}, G_q, and G_{13}), or the peculiar G_{15}, induced heterotrimeric G protein dissociation and a quantifiable reduction in BRET signal. Applying this method, the authors dissected the coupling profile of 22 class A oGPCRs out the 48 analyzed.[36] A different BRET assay was also developed by Schihada and colleagues for the same purpose.[37] In this assay, a tricistronic plasmid encoded different Gα proteins fused with NanoLuc in an internal site, together with a Gβ protein and a Gγ protein fused N-terminally with cpVenus. With such a configuration, GPCR activation results in a reduction in BRET signal that was sensitive enough to detect also constitutive activity. To this aim, the authors analyzed the G protein coupling profile of 3 oGPCRs (GPR3, GPR6, and GPR12) and found G_s coupling for all of them and G_o coupling only for GPR12.[37]

A different strategy aimed at defining the G protein coupling profile of oGPCRs has been developed engineering a library of chimeric GPCRs that bear the signaling domains of oGPCRs and the light sensitive domain of rhodopsin.[39] In detail, the authors of this study created a library of constructs for mammalian expression of human GPCR chimeras where they swapped the intracellular loops and the C-terminal region of rhodopsin with the corresponding domains of each of 74 GPCRs of interest, including 63 class A oGPCRs. The effects of light-induced stimulation of these chimeric oGPCRs in transfected cells were then quantified using luciferase reporter assays as described above. The results of this elegant study revealed the G protein coupling profile of 15 oGPCRs.

Overall, studies aimed at defining the G protein coupling profile or elucidating the intracellular signaling cascades activated by oGPCRs are valuable because of at least two reasons: (1) they provide key information about the biology of oGPCRs enabling studies about their pathophysiological role; and (2) they provide a basis to develop targeted strategies for deorphanization efforts or for identification of synthetic compounds modulating oGPCR activity that are much needed tools for pharmacological research. On the other side, an obvious limitation of studies aiming at recreating oGPCR signaling in heterologous systems is the lack of positive controls. In fact, GPCRs overexpressed in cells may not always be functional as they sometimes require accessory proteins to fold properly, to traffic to the plasma membrane, or partnering with other GPCRs to form functional heteromers. Furthermore, studies designed to measure constitutive activity rely on high GPCR activity in the absence of endogenous ligands, limiting the success of this approach to a subset of receptors.

3. Universal platforms to detect oGPCR activation for high-throughput screening purposes

Deorphanization strategies, intended as methods to identify endogenous ligands activating oGPCRs (Table 1), are frequently supported by parallel efforts aimed at isolating synthetic ligands (Table 2). In fact, the identification of agonists, antagonists, or inverse agonists facilitates pharmacological studies on oGPCRs that often lead to a greater understanding of the pathophysiological role played by oGPCRs in relevant systems. Knowledge about GPCR mechanisms of activation and coupling to intracellular signaling pathways enables the development of high-throughput screening platforms suitable to identify both endogenous and synthetic ligand-receptor pairs. In fact, screening of large libraries of compounds can only be successful when the right intracellular output modulated by the oGPCR is measured. In the past few decades, the lack of data about signaling cascades activated by oGPCRs prompted the search for universal signaling mechanisms activated by the majority of the GPCRs.

Table 1 Recently identified candidate ligand-receptor pairs.

Receptor	Endogenous ligands	Ligand identification strategy	Assay validation	References
GPR1	Osteocrin Gastrin releasing peptide Cholecystokinin	Sequence alignment for peptidic precursors, evolutionary conserved domains and ligand pocket characterization	β–Arrestin recruitment	27
GPR15	Protein GPR15L	Sequence alignment for peptidic precursors, evolutionary conserved domains and ligand pocket characterization	DMR, internalization, cAMP, β-arrestin recruitment	27
GPR31	Lactic acid Piruvic Acid	CoiN-pocket workflow (ligand pocket alignment and score for receptor–ligand interactions)	cAMP	40
GPR35	Kinurenic acid	Compound screening	Ca^{2+} mobilization, IP1	41
GPR55	MANSC domain-containing peptie 1 PACAP-27 Sperm associated antgen 11B Secretogranin-1 β-microseminoprotein Clusterin-like protein 1	Sequence alignment for peptidic precursors, evolutionary conserved domains and ligand pocket characterization	DMR, internalization	27
GPR68	Osteocrin CARTp Pro-opiomelanocortin	Sequence alignment for peptidic precursors, evolutionary conserved domains and ligand pocket characterization	DMR, internalization, Ca^{2+} mobilization, cAMP	27

Receptor	Ligand	Method	Assay	Reference
GPR75	20-HETE	Click chemistry derivative of 20-HETE suitable for crosslinking	IP1, radioligand binding	42
GPR83	PEN	Expression correlation in brain regions for ligand and receptor	Ca^{2+} mobilization	43
GPR101	RvD5	Screening of oGPCRs potentially mediating RvD5 effects	β-Arrestin recruitment, DMR	44
GPR139	L-Trp L-Phe ACTH α-MSH β-MSH HRFW domain	Computational approaches	Ca^{2+} mobilization, cAMP	45,46
GPR146	C-peptide	Testing the active compound on cell lines, cross evaluation of oGPCRs expressed	Reporter induction	47
GPR158	Osteocalcin	Expression correlation in brain regions for ligand and receptor	co-IP with ligand and IP1	48
GPR171	BigLEN	Expression correlation in brain regions for ligand and receptor	Ca^{2+} mobilization	49
GPR182	Chemokines (CXCL10, CXCL12, CXCL13)	Closest receptor paralogue is a chemokine receptor	Ca^{2+} mobilization, NanoBit Technology	50

Continued

Table 1 Recently identified candidate ligand-receptor pairs.—cont'd

Receptor	Endogenous ligands	Ligand identification strategy	Assay validation	References
BB3	Neuromedin–U	Sequence alignment for peptidic precursors, evolutionary conserved domains and ligand pocket characterization	DMR, internalization, IP1, β-arrestin recruitment	27
	Neuromedin–B			
	Proenkephaline-A			
	Gastrin Releasin peptide-1			
LRG4	R-spondin	Correlation with Frizzled receptor signaling and immunoprecipitation	co-IP with ligand and TopFLASH system reporter for Wnt signaling	51,52
LRG5	R-spondin	Correlation with Frizzled receptor signaling and immunoprecipitation	co-IP with ligand and TopFLASH system reporter for Wnt signaling	51,52
GPR160	CARTp	High correlation between tissue expression profile of GPR160 and CARTp	*in vitro* assays for gene expression and protein phosphorylation	53
P2Y10	LysoPS	From previously suggested ligand	IP1, RhoA activation, cAMP	54
	ATP	Structural homolog to Purinergic receptor	Ca^{2+} mobilization	
MRGPRX1	LVV–7 / VV–7	Fractions obtained from HPLC of biological samples, MALDI-TOF	Ca^{2+} mobilization	55
MRGPRX4	Bile Acids	Correlation between bile acids pathological conditions and agonist side effects	Ca^{2+} mobilization, TGF-α shedding	56,57

Table 2 Recently published synthetic agonists for oGpCRs.

Receptor	Synthetic ligands	Ligand identification strategy	Assay validation	References
GPR18	Tricyclic xantines	Compound screening	β-Arrestin recruitment	58
GPR27	PubChem CID: 175606 and 1,177,181	Compound screening	β-Arrestin recruitment, TGF-α shedding, cAMP, Ca^{2+} mobilization	59
GPR35	Cromolyn	Compound screening	β-Arrestin recruitment	60,61
	Lodozamide	Structural similarities to Cromolyn		
	Bufrolin			
GPR139	LP-360924	Compound screening	Ca^{2+} mobilization	62
GPRC5A	TA and 7FTA	Chemoproteomics with indole-3-acetic acid derivatives	β-Arrestin recruitment	63

These molecular signatures have become the foundation of screening methods developed in the past years; concepts and design of some of these platforms are briefly summarized and discussed here.

3.1 Calcium mobilization

Calcium mobilization is a common downstream effect of $G\alpha_q$-coupled receptor activity. The observation that $G\alpha_q$ family members $G\alpha_{15}$ or $G\alpha_{16}$ show promiscuous coupling to GPCRs normally activating other G protein families enabled the implementation of this widely used assay, where receptor activation was redirect to calcium release from intracellular stores and detected through calcium-binding fluorescent probes.[64,65] Similarly, libraries of Gq-based Gα protein chimeras where the C-terminal domains were swapped with those of each of the other Gα proteins have been used in combination with calcium detection to measure GPCR activation.[66,67]

3.2 TGF-α shedding assay

Signaling cascades activated by $G_{q/11}$ and $G_{12/13}$ protein subfamilies have been recently associated with increased activity of the transmembrane enzyme A Disintegrin And Metalloprotease 17 (ADAM17).[21] More in details, ADAM17 was found able to cleave a set of substrates including the Tumor Growth Factor α (TGF-α). This peculiar signaling was later exploited by Inoue and colleagues to design a high-throughput screening platform named TGF-α shedding assay.[21,35] Here, TGF-α was fused to the enzyme alkaline phosphatase (AP) so that ADAM17 cleavage of TGF-α could be quantified as AP activity in the collected extracellular medium. To make this method applicable to different GPCRs, the authors took advantage of Gα protein chimeras bearing the core of $G\alpha_q$ and swapping the C-terminus with those of every other Gα protein. This configuration made it is possible to detect receptor activation when the correct Gq-chimera is co-transfected. However, a significant fraction of the receptors tested appeared to positively couple to many of the chimeras tested indicating a certain level of promiscuity. This could be due to different factors along the signaling pathway, from the Gq-chimeras to alternative mechanisms regulating ADAM17 activity. Nonetheless, the TGF-α shedding assay represents a powerful approach as it can detect activation of many receptors including oGPCRs.[21]

3.3 [35S]GTPγS binding assay

An *in vitro* assay that can be considered universal as it relies on a common feature of GPCR activation consists in the measurement of GPCR signaling *via* radio-labeled GTP binding, or [35S]GTPγS binding assay. Here, the levels of G protein activation following agonist application are measured as an increased retention of radioactive GTP to isolated cell membranes.[68] In spite of the obvious advantage of being a universal readout of G protein activation at the most proximal site, this strategy is not ideal to "scale up" to the high-throughput level required for identification of oGPCR ligands.[69]

3.4 β-Arrestin recruitment

In receptor pharmacology, signal activation is commonly coupled to desensitization and endocytosis of the receptor in order to terminate the cellular response. Deep understanding of these orchestrated mechanisms represents a valuable source for developing cell-based assays. In particular, after GPCR activation, several kinases phosphorylate the C-terminus of the receptor promoting the recruitment of β-arrestins to the GPCR and blocking Gα protein re-coupling, allowing receptor desensitization and then endocytosis.[70] Therefore, screens for receptor activation based on β-arrestin recruitment are believed to represent a universal deorphanization strategy. Assays aiming at detecting β-arrestin recruitment often involve heterologous expression of receptors and/or β-arrestin fused to tags or enzymes that allow the detection of receptor-arrestin complex formation using a variety of readouts. The number of β-arrestin assays is large and includes PathHunter (DiscoverX),[71] Tango GPCR Assay (Thermo Fisher Scientific),[72] PRESTO-Tango,[73] LinkLight GPCR/ β-arrestin Signaling Pathway Assay (BioInvenu),[74] and Transfluor (Molecular Devices).[75] All these assays are suitable for high-throughput screening and the biophysical concepts underlying some of these platforms are summarized below. In the PathHunter assay, a β-galactosidase is splitted between the receptor and β-arrestin. After β-arrestin recruitment, the functional enzyme is reconstituted by complementation of the two parts and a chemiluminescent signal is generated by addition of an exogenous substrate.[71] In the Tango assay a transcription factor is fused to the receptor intracellular C-terminus, preceded by a Tobacco Etch Virus (TEV) protease specific cleavage site, while β-arrestin is associated with TEV. After β-arrestin recruitment, the transcription factor is released and expression of a reporter gene is induced.[72,76] More recently, a novel version of this assay named PRESTO-Tango has been developed where the expressed GPCRs were tagged in the N-terminal region with a signal peptide promoting plasma membrane localization, while a V2 tag was placed in the C-terminal tail.[73] Interestingly, the insertion of the V2 tag in the C-terminus of GPR27 in combination with enzyme complementation of NanoLuc (NanoBiT) was used for a similar β-arrestin recruitment assay applied to the identification of ligands of GPR27.[59] The NanoBiT technology is widely used to investigate protein-protein interactions detected as reconstitution of a splitted reporter luciferase made of a large fragment (lgBiT) and small fragment (smBiT).[77] The LinkLight technology allows the detection of specific protein-protein interactions and can be therefore exploited to identify interaction of signaling proteins, among them GPCR-arrestins.[74] The main characteristic of LinkLight is the use of a permutated luciferase (pLuc) that is activated upon cleavage by a TEV protease. By inserting the pLuc and TEV on the two proteins of interest it is possible to detect a bioluminescence signal after supplying the luciferase substrate if the interaction occurs. Advantages of this method are the specificity of the signal and low background in response to the stimulus. However, the proteins of interest still need to be tagged, potentially altering their protein-protein interaction. The Transfluor assay is based on a fluorescently tagged β-arrestin which after receptor activation is redistributed within few seconds at the plasma membrane, and after minutes in endocytic vesicles.[75] The Transfluor assay works with both confocal and non-confocal imaging, however confocal systems allow for a more precise detection of weak signals.[78] This assay is suitable for high-throughput screening of compounds activating GPCRs and allows to monitor the interplay between receptors and β-arrestin with spatio-temporal resolution when combined with imaging systems for automated analysis such as the INCELL Analyzer, Arrayscan, and ImageeXpress.[75,78–81]

3.5 Enhanced bystander BRET (ebBRET)

Assays able to detect protein-protein interaction such as BRET are widely used and versatile methods for GPCR drug discovery as they enable the quantification of specific interactions and dissociations in a time resolved manner.[82,83] Namkung and colleagues previously developed three BRET sensor pairs allowing to monitor GPCR activation and subsequent endocytosis, implementing this technique by using RLucII and GFP, which do not interact spontaneously, and thus reducing non-specific signals (enhanced bystander BRET or ebBRET). In the first pair, the receptor is tagged at the C-terminus with RlucII, while the donor is a GFP fused with a domain from the Lyn kinase, responsible for the targeting of the protein to the plasma membrane. Upon receptor activation, a reduction in BRET signal occurs, due to translocation of the receptor to intracellular vesicles and subsequent inability to excite the fluorophore. The second pair of sensors is based on luciferase tagged receptors and GFP fused to the FYVE domain of endofin (rGFP) that targets GFP exclusively to endocytic vesicles. Upon receptor activation and endocytosis, an increase in BRET signal is detected. The third pair of biosensors is represented by β-arrestin-RlucII and rGFP, which leads to BRET signal only after receptor endocytosis. The use of BRET-based sensors allows to follow the dynamic of the interaction in live cells, however it is not ideal for high-throughput formats.[82] Based on this work, Pedersen and colleagues implemented a novel approach switching the modality of signal detection from BRET to a protein complementation, and named these systems MeNArc and EeNArc. These approaches have not been yet applied to oGPCR screening, and are based on recruitment of β-arrestin to GPCRs at the plasma membrane (MeNArc) and early endosomes (EeNArc). In particular, the MeNArc and EeNArc assays rely on the N-terminal fragment (1–102 aa) of NanoLuc anchored to the plasma membrane through the double palmitoylated domain of GAP43, or to endocytic vesicles through the FYVE domain of endofin, respectively. The C-terminal fragment (103–172 aa) of NanoLuc is fused to β-arrestin, so that after receptor activation and β-arrestin recruitment, the enzyme is complementary assembled and able to produce bioluminescence when a substrate is applied.[84] This technology seems promising, since the receptor is not modified with tags that could hamper its ligand binding or signaling.

3.6 Real-time internalization assay

The Real-Time internalization assay is an alternative method developed to deorphanize oGPCRs based on Fluorescence Resonance Energy Transfer (FRET) between a terbium labeled SNAP tag fused on extracellular domains of the receptor of interest, and fluorescein in the extracellular media. In this assay, receptor activation and subsequent internalization are detected as a decrease in FRET signal.[85]

3.7 Dynamic mass redistribution (DMR) assay

As previously mentioned, the assays described are limited by the use of modified receptor, G proteins, or β-arrestin that could negatively impact receptor activation and signaling properties. Therefore, efforts to develop a universal platform suitable for every GPCR overcoming these problematics are still ongoing. To this goal, a completely different strategy has been generated with the goal of quantifying GPCR activation by measuring alterations of the cytoskeleton, commonly referred to as Dynamic Mass Redistribution (DMR) assay.[86] Here, components of the cytoskeleton are reorganized by G protein activation, thus inducing changes of impedance that can be monitored using technologies such as the CellKey system. This approach can be implemented in different cell lines or primary cells, which represents another advantage compared to the most common screening assay that are normally carried out in cell lines chosen for their predisposition to be easily transfected. However, label free systems may result in outcomes deriving from simultaneous activation of multiple pathways, leading to false negative or false positive results.[86,87] Noteworthy, DMR was especially useful in the identification of peptidic ligands for GPR1, GPR15, GPR55, GPR68 and BB$_3$.[27]

3.8 Ligand-induced forward trafficking (LIFT) assay

Screening assays for GPCRs usually rely on expression of tagged versions of the receptor; similarly mutated receptors can provide a convenient setup in relevant assays. The recently developed LIFT assay is based on the recruitment of an engineered target receptor from the endoplasmic reticulum to the plasma membrane upon ligand interaction.[88] More in details, the receptor is genetically modified with mutations enabling ER retention and it is fused with a β-galactosidase fragment which will reconstitute the fully active enzyme in case of positive probing. After the target GPCR reaches the plasma membrane, it is internalized in endosomes, where β-galactosidase is reconstituted allowing to detect the ligand-receptor interaction as a chemiluminescent signal. This approach was successfully applied to isolate synthetic ligands for the orphan GPR20. A disadvantage of this assay is its applicability only for ligands that are able to cross the plasma membrane and it does not detect specific agonists but any type of interacting molecules.[88]

3.9 RNA sequencing of barcoded reporters

An exception to the search for universal platforms is represented by screening efforts for olfactory GPCRs. Olfactory receptors represent the largest subfamily of GPCRs in the human genome, and therefore screening platforms to identify ligand-receptor pairs require a fast readout on large scale analysis. Even though signaling pathways activated by most oGPCRs are not clear, for olfactory GPCRs it is established that they transduce signals through G_{olf} activation, which is a G_s family member enriched in the olfactory bulb. Therefore, in this specific case, reporter systems downstream of $G\alpha_{s/olf}$ protein activation can be applied to high-throughput screening for the identification of novel ligand-receptor pairs. Accordingly, Jones and colleagues developed a multiplex assay with several stable cell lines each expressing a specific olfactory receptor and a reporter tagged with a 15 nucleotide barcode.[89] Each olfactory receptor was expressed under an inducible promoter responsive to tetracycline (TRE) and displayed a tag in the N-terminal region to verify its plasma membrane localization, while the barcode was expressed under the control of a CRE promoter. The same cell line also stably expressed $G\alpha_{olf}$ and chaperone proteins relevant for receptor signaling and membrane localization. Activation of the receptor initiated G_{olf} signaling prompting cAMP production and inducing the CRE system and barcode transcription. The use of RNA-seq finally allowed the detection of receptor activation. This system has the advantage of combining different stable cell lines in a single well, making therefore one sample sufficient for the screening of many olfactory Gpcrs as demonstrated by the successful deorphanization of 15 olfactory receptors.[89] Disadvantages of this system are the high costs of RNA sequencing and the need to maintain many stable cell lines.

Limitations of individual assays used for ligand discovery have been driving research aimed at improving high-throughput screening platforms for oGpcrs. These issues emerged clearly when alternative strategies are used in parallel to test or to validate receptor-ligand pairs.[27] Thus, even if time-consuming and more expensive, adopting multiple assays utilizing different readouts will likely ensure a greater success in the identification of ligand-receptor pairs than individual assays.

4. Targeted deorphanization strategies

An effective approach to generate hypothesis about pairing of endogenous ligands with oGpcrs consists in identifying overlapping localization *in vivo*. Historically, this was achieved by the application of radiolabeled ligands on tissues and organ slices and matching these studies with thorough analysis of oGPCR expression by *in situ* hybridization experiments or, when suitable antibodies were available, immunohistochemistry.[49] More recently, single cell RNA sequencing techniques enabled researchers to enhance the resolution of this approach through access to databases that can be interrogated for cellular specificity of receptor expression.[90–93] This strategy can narrow down the number of candidate ligands that can foreseeably interact with oGpcrs specifically localized in certain regions. Here, we describe a few successful examples of deorphanization achieved in this way.

4.1 GPR171-BigLEN

The identification of GPR171 as the receptor of the peptide BigLEN followed this powerful approach by merging functional and expression information from different sources.[49] In detail, the authors first characterized in which rat body region a radiolabeled BigLEN peptide was accumulated, finding it mostly enriched in the hypothalamus. Then, they observed that BigLEN induced cellular signaling mediated by a GPCR, as it increased [^{35}S]GTPγS binding. They also showed that the readout of this assay was pertussis toxin sensitive, therefore pointing at a $G_{i/o/z}$-coupled receptor. Finally, they found that Neuro2A cell lines were responsive to BigLEN suggesting endogenous expression of its receptor. At this point, by comparing GPCR expression databases for oGPCRs that were enriched in both hypothalamus and Neuro2A cells, they obtained a shortlist of only four potential target receptors, among them GPR171. Subsequent functional studies with a variety of screening platforms to explore the activation of these oGPCRs led to the discovery of GPR171 as the receptor for BigLEN. Following a similar pipeline, the same research group discovered the neuropeptide PEN as a ligand for GPR83.[43]

GPR160-CARTp. An analogous approach has been recently applied to identify the endogenous ligand for GPR160 predicted to be of peptidic nature. Here, a cross evaluation for tissue expression profile of GPR160 and a subset of candidate peptides resulted in a high correlation between GPR160 and the cocaine- and amphetamine- regulated transcript peptide (CARTp).[53] Yosten and colleagues were able to confirm both CARTp-mediated signaling as dependent on GPR160 expression, as well as CARTp-GPR160 interaction through proximity ligation assay.[53] Moreover, they revealed that GPR160 mediated the effects of CARTp on neuropathic pain *in vivo*, shedding light on a new pharmacological target for this pathological condition. Later, the same group reported that CARTp also modulates food and water intake *via* activation of GPR160.[94]

GPR139- L-Trp and GPR139-L-Phe. Research on GPR139 showcases how the use of complementary approaches can lead to identification of endogenous ligands for oGPCRs. A sequence alignment of conserved residues among human expressed GPCRs pointed at class A GPR139 as a homolog of the thyrotropin-releasing hormone receptor (TRHR).[95] Furthermore, the presence of GPR139 endogenous ligands in brain extract was demonstrated using a $G\alpha_{16}$-based calcium mobilization assay.[96] By combining these information, scientists searched for possible peptidic ligands for this receptor in the central nervous system. First, a high-throughput screening based on cAMP detection led to the identification of the first set of synthetic agonists for GPR139.[62] Later, another group outlined the structure-activity relationship (SAR) of GPR139 ligands; this information was then used to develop a pharmacophore model as a computational approach for screening of possible GPR139 ligands.[45] Application of this tool resulted in the identification of L-tryptophan and *L*-phenylalanine as endogenous GPR139 ligands, which was then confirmed using several assays including [^{35}S]GTPγS binding, calcium mobilization, and ERK phosphorylation.[97] More recently, the alignment of the ligand binding pocket of GPR139 and other class A receptors, suggested proopiomelanocortin derived peptides ACTH, α-MSH, β-MSH as possible GPR139 ligands. These hormones and their conserved domain HRFW were found able to induce GPR139-mediated calcium mobilization and cAMP production.[46]

GPR158-Osteocalcin. A targeted strategy was recently applied to identify the oGPCR mediating the effects of osteocalcin in the central nervous system.[48] The bone-derived hormone osteocalcin was previously shown to activate the class C receptor GPRC6A,[98,99] even though other groups failed to reproduce this functional interaction.[100,101] Since GPRC6A is not expressed in the brain, a search for an alternative class C GPCR that could be possibly activated by osteocalcin was conducted. This hunt led to the identification of GPR158.[48] Co-immunoprecipitation experiments followed by functional studies showing that osteocalcin-mediated effects were ablated in GPR158 KO animals seemed to support a role for GPR158 as a receptor for osteocalcin. However, adaptation events leading to the alteration of a variety of signaling pathways observed in GPR158 KO mice[5] represent a valid alternative explanation to the observed outcome. Validation of this ligand-receptor pair by other groups, possibly in a reconstituted system will clarify if GPR158 could be effectively considered deorphanized.

The described examples represent just a fraction of the recent successful deorphanization studies. Analogue approaches based on evaluation of active compound effects on biological systems followed by searches to narrow down the number of possible oGPCR targets recently led to the identification of candidate ligands for GPR101,[44] MRGPRX4,[56,57] and GPR146.[47] Targeted approaches are low-throughput by definition as they rely on peculiar features of either the oGPCR of interest or its endogenous candidate ligand. Nonetheless, taking advantage of the constantly growing amount of data generated by –omics studies, these strategies are likely the most powerful resource leading to groundbreaking discoveries in the oGPCR field. Such a holistic approach in examining oGPCRs is also what makes these discoveries more relevant from a pathophysiological and a pharmacological perspective.

5. Proteomics and chemoproteomics

An effective strategy to identify peptides and proteins as endogenous ligands for oGPCRs consists in isolating receptors from native tissues and apply mass spectrometry to reveal extracellular binding partners.[102] This method is especially suitable for oGPCRs that are predicted to interact with partners of peptidic nature, when antibodies against the receptor appropriate for immunoprecipitation are available. Alternatively, knock-in animal models where affinity tags are fused with the oGPCR expressed *in vivo*, or the use of Fc-fragments fused with extracellular domains of the oGPCR as bait, can make up for this lack of tools.

This approach was successfully employed by de Lau and colleagues in their research focused on the oGPCRs LGR4 and LGR5.[51] They first found that interfering with LGR4 and LGR5 expression in crypt cells of the intestine led to a reduction in stem cell proliferation, an event that is normally mediated by frizzled receptors through the Wnt-signaling pathway. Then, they investigated a possible relation between frizzled receptors and LGRs by modulating frizzled receptor activation, discovering an interplay between these molecular components. To explore this relationship, they generated tagged baits of LGRs and through mass spectrometry they identified frizzled receptors as binding partners. This discovery led to the hypothesis of a possible interaction between LGRs and the frizzled receptor agonists R-spondins. To test this hypothesis, the authors exploited tagged-ectodomains of LGRs which co-immunoprecipitated with R-spondin-Fc, suggesting that LGRs are part of frizzled receptor complexes able to interact with R-spondins.[51] Finally, they showed that LGRs act as additional Wnt receptor components mediating Wnt signal enhancement by soluble R-spondins. Immediately after the publication of these results, another group confirmed R-spondins as ligands for LGR4 and LGR5.[52]

More recently, in an effort to isolate extracellular binding partners, including possible endogenous ligands for GPR158 and its homolog GPR179, a proteomics approach was successfully applied.[103,104] These oGPCRs display peculiar extracellular regions with EGF-like calcium-binding domains usually associated with extracellular matrix interactions. Using ectodomains of GPR158 and GPR179 fused to Fc fragments as bait, interactions with several members of the Heparan Sulfate Proteoglycan (HSPG) family, a heterogeneous group of extracellular matrix components, were identified. HSPGs are characterized by the presence of at least one covalently attached heparan sulfate (HS) chain, a type of glycosaminoglycan. Further investigations established that these interactions are required to form and maintain stable trans-synaptic complexes essential to achieve normal neurotransmission both in the retina (GPR179-Pikachurin)[104] and in the hippocampus (GPR158-Glypican4).[103] The possible role of HSPGs as endogenous candidate ligands for GPR158 and GPR179 remains to be examined.

Chemical proteomics is an innovative strategy recently developed to identify binding sites of small molecules on target receptors.[105,106] Chemoproteomics relies on the chemical modification of ligands bearing the potential to activate endogenous oGPCRs and their use as baits for proteomics studies with the goal of identifying their interactome, possibly an oGPCR. The chemical design of probes suitable for chemoproteomics requires the presence of a UV light photoreactive cross-linker group, a biotin subunit, and a chemical scaffold with affinity for the target GPCR. Moreover, these probes need to maintain their affinity to the target GPCR after the required chemical modifications. In the past, chemoproteomics approaches applied to study GPCRs have been limited to the identification of off-targets for 5-HT1A and 5-HT6 receptors ligands.[107] More recently, Zhao and colleagues applied a similar approach to deorphanize GPRC5A, an understudied class C oGPCR.[63] In this study, the authors investigated microbiota-derived aromatic compounds that have been suggested as candidate ligands for GPCRs. Specifically, they derivatized indole-3-acetic acid (IAA), a common microbiota-generated metabolite, with bio-orthogonal tags for detection and affinity enrichment, and with a photo-affinity moiety for UV crosslinking. These new chemicals where then used as baits to isolate IAA-interacting proteins in intestinal epithelial cells that were later identified by mass spectrometry. This approach led to the identification of GPRC5A among a few other GPCRs as a candidate receptor for IAA. To test whether this interaction was producing GPRC5A activation, they used the PRESTO-Tango assay and treated cells expressing GPRC5A with a library of compounds including IAA and molecules sharing a similar chemical profile. Interestingly, they found that despite the interaction between IAA and GPRC5A, this did not generate any detectable signal, while they observed a dose-dependent activation of GPRC5A by similar compounds tested, namely tryptamine (TA) and its derivative 7-fluoro-tryptamine (7FTA). Further studies using orthogonal methods to test GPRC5A activation by TA and 7FTA will help to clarify mechanisms of ligand-receptor activation and define whether any of these compounds could indeed be the endogenous ligand for GPRC5A.

Chemoproteomics strategies were also recently applied to the identification of the endogenous GPCR mediating the biological effects of 20-hydroxyeicosatetraenoic acid (20-HETE).[42] Chemical compounds suitable for click-chemistry, crosslinking through UV light, and analog to the active endogenous ligand 20-HETE were generated. After compound application to cultured human endothelial cells and UV crosslinking,

GPR75 was identified as the binding partner for 20-HETE by proteomics. These promising results indicate that chemoproteomics approaches are valuable new tools in the deorphanization of oGPCRs.

6. Computational methods

In silico studies have recently revolutionized the oGPCR field by enabling the cross analysis of large scale datasets opening the door to previously inaccessible information. Applying technically advanced machine learning approaches is now possible to create simulations of ligand binding at GPCRs as well as deeply explore increasingly detailed databases. Computational approaches with conserved evolutionary domain analysis, for example by using resources like Eukaryotic Linear Motif (ELM), often allow to identify amino acid sequences relevant for protein function or protein-protein interaction. When considering oGPCRs, this can be performed both for peptidic ligands and for receptors, as recently illustrated.[27] Applying this strategy, Foster and colleagues first searched the whole human proteome for novel precursors of potentially released peptides by exploiting knowledge about the secretion motifs of known peptides. Then, they analyzed the presence of evolutionary conserved sequences and found that peptide coding regions, including the dibasic cleavage sites, are more conserved than other parts of the precursor. Accordingly, based on the most conserved sequences potentially encoding for active peptides, they generated a library of 218 compounds. Guided by structural distinct features identified in well studied peptide receptors, they generated a list of 21 class A oGPCRs as potential receptors for peptides. The 218 novel peptides were then applied on cells expressing each of the shortlisted 21 oGPCRs and activation of cellular signaling was measured using several platforms including DMR, PRESTO-Tango assay, Real-Time receptor internalization, second messenger measurements (IP1 and cAMP), and β-arrestin recruitment (PathHunter). This powerful, multifaceted approach was successful in identifying novel endogenous ligands for oGPCRs BB₃, GPR1, GPR15, GPR55, and GPR68 expanding our knowledge of the human signaling systems.

Phylogenetic similarities among the ligand binding pocket of GPCRs have not always proved useful in predicting GPCR–ligand interactions. To improve this approach, a novel computational method that takes advantage of growing data about small ligand binding sites in GPCR structures and that ranks the importance of individual residue–ligand interactions was developed.[108,109] The goal of this method, dubbed GPCR–CoINPocket, is the identification of surrogate ligands. Even if the original paper describing this method was retracted due to technical issues related to the validation of a predicted GPR37L1 ligand, GPCR-CoINPocket still represents an innovative tool in oGPCR deorphanization. In fact, a predicted GPCR-ligand pair for GPR31 was validated in subsequent independent studies.[40]

Finally, the cryo-EM revolution of the last few years will foreseeably boost future computational studies on oGPCRs using molecular docking. In molecular docking strategies, the chemical structure of the target receptor is used to generate computational simulations to fit libraries of chemical compounds in the GPCR orthosteric binding pocket.[110,111] The potential of rapidly testing millions of compounds for their ability to interact with a receptor, in this case oGPCRs, is only limited by the number of available structures to implement this strategy. In fact, currently only three class A and one class C oGPCR structures have been solved.[112–117] However, the constant improvement of cryo-EM technologies is driving a fast growing increase in the number of solved GPCR structures that will likely soon include many more oGPCRs. It is conceivable that new structures will provide tools for a more widespread application of molecular docking facilitating future deorphanization efforts.

7. Conclusions and future directions

oGPCRs are an underutilized reservoir of potential drug targets. The development of novel tools to explore their signaling properties, biology, and ligand identification is therefore fundamental in the process of enriching our arsenal of therapeutic drugs. Technologies and strategies described above represent only a fraction of the recent innovations aimed at deorphanizing and understanding oGPCR physiology. Improvements in the design of cell-based assays together with the generation of novel more sensitive biosensors for intracellular detection of GPCR activation will likely accelerate the deorphanization process. Nonetheless, future discoveries in the oGPCR field will most likely depend on the availability of detailed oGPCR structures to guide hypothesis related to the nature of their endogenous ligands.

References

1. Sriram K, Insel PA. G protein-coupled receptors as targets for approved drugs: How many targets and how many drugs? *Mol Pharmacol*. 2018;93(4):251–258. https://doi.org/10.1124/mol.117.111062.
2. Audo I, Bujakowska K, Orhan E, et al. Whole-exome sequencing identifies mutations in GPR179 leading to autosomal-recessive complete congenital stationary night blindness. *Am J Hum Genet*. 2012;90(2):321–330. https://doi.org/10.1016/j.ajhg.2011.12.007.

3. Peachey NS, Ray TA, Florijn R, et al. GPR179 is required for depolarizing bipolar cell function and is mutated in autosomal-recessive complete congenital stationary night blindness. *Am J Hum Genet.* 2012;90(2):331–339. https://doi.org/10.1016/j.ajhg.2011.12.006.

4. Smith EL, Harrington K, Staehr M, et al. GPRC5D is a target for the immunotherapy of multiple myeloma with rationally designed CAR T cells. *Sci Transl Med.* 2019;11 (485). https://doi.org/10.1126/scitranslmed.aau7746.

5. Sutton LP, Orlandi C, Song C, et al. Orphan receptor GPR158 controls stress-induced depression. *Elife.* 2018;(7). https://doi.org/10.7554/eLife.33273.

6. Watkins LR, Orlandi C. Orphan G protein coupled receptors in affective disorders. *Genes (Basel).* 2020;11(6). https://doi.org/10.3390/genes11060694.

7. Rodgers G, Austin C, Anderson J, et al. Glimmers in illuminating the druggable genome. *Nat Rev Drug Discov.* 2018;17(5):301–302. https://doi.org/10.1038/nrd.2017.252.

8. Oprea TI, Jan L, Johnson GL, et al. Far away from the lamppost. *PLoS Biol.* 2018;16 (12):e3000067. https://doi.org/10.1371/journal.pbio.3000067.

9. Stoeger T, Gerlach M, Morimoto RI, Nunes Amaral LA. Large-scale investigation of the reasons why potentially important genes are ignored. *PLoS Biol.* 2018;16(9): e2006643. https://doi.org/10.1371/journal.pbio.2006643.

10. Oprea TI, Bologa CG, Brunak S, et al. Unexplored therapeutic opportunities in the human genome. *Nat Rev Drug Discov.* 2018;17(5):317–332. https://doi.org/10.1038/nrd.2018.14.

11. Davenport AP, Alexander SP, Sharman JL, et al. International Union of Basic and Clinical Pharmacology. LXXXVIII. G protein-coupled receptor list: Recommendations for new pairings with cognate ligands. *Pharmacol Rev.* 2013; 65(3):967–986. https://doi.org/10.1124/pr.112.007179.

12. Alexander SP, Christopoulos A, Davenport AP, et al. The concise guide to pharmacology 2021/22: G protein-coupled receptors. *Br J Pharmacol.* 2021;178(Suppl 1): S27–S156. https://doi.org/10.1111/bph.15538.

13. Zhang X, Mantas I, Fridjonsdottir E, Andren PE, Chergui K, Svenningsson P. Deficits in motor performance, neurotransmitters and synaptic plasticity in elderly and experimental parkinsonian mice lacking GPR37. *Front Aging Neurosci.* 2020;12:84. https://doi.org/10.3389/fnagi.2020.00084.

14. Meirsman AC, Le Merrer J, Pellissier LP, et al. Mice lacking GPR88 show motor deficit, improved spatial learning, and Low anxiety reversed by Delta opioid antagonist. *Biol Psychiatry.* 2016;79(11):917–927. https://doi.org/10.1016/j.biopsych.2015.05.020.

15. Oishi A, Karamitri A, Gerbier R, Lahuna O, Ahmad R, Jockers R. Orphan GPR61, GPR62 and GPR135 receptors and the melatonin MT2 receptor reciprocally modulate their signaling functions. *Sci Rep.* 2017;7(1):8990. https://doi.org/10.1038/s41598-017-08996-7.

16. Ahmad R, Wojciech S, Jockers R. Hunting for the function of orphan GPCRs - beyond the search for the endogenous ligand. *Br J Pharmacol.* 2015;172 (13):3212–3228. https://doi.org/10.1111/bph.12942.

17. Levoye A, Dam J, Ayoub MA, Guillaume JL, Jockers R. Do orphan G-protein-coupled receptors have ligand-independent functions? New insights from receptor heterodimers. *EMBO Rep.* 2006;7(11):1094–1098. https://doi.org/10.1038/sj.embor.7400838.

18. Nygaard R, Zou Y, Dror RO, et al. The dynamic process of beta(2)-adrenergic receptor activation. *Cell.* 2013;152(3):532–542. https://doi.org/10.1016/j.cell.2013.01.008.

19. Corder G, Doolen S, Donahue RR, et al. Constitutive mu-opioid receptor activity leads to long-term endogenous analgesia and dependence. *Science.* 2013;341 (6152):1394–1399. https://doi.org/10.1126/science.1239403.

20. Damian M, Marie J, Leyris JP, et al. High constitutive activity is an intrinsic feature of ghrelin receptor protein: A study with a functional monomeric GHS-R1a receptor reconstituted in lipid discs. *J Biol Chem*. 2012;287(6):3630–3641. https://doi.org/10.1074/jbc.M111.288324.

21. Inoue A, Ishiguro J, Kitamura H, et al. TGFalpha shedding assay: An accurate and versatile method for detecting GPCR activation. *Nat Methods*. 2012;9(10):1021–1029. https://doi.org/10.1038/nmeth.2172.

22. Rosenbaum DM, Rasmussen SG, Kobilka BK. The structure and function of G-protein-coupled receptors. *Nature*. 2009;459(7245):356–363. https://doi.org/10.1038/nature08144.

23. Mathiasen S, Palmisano T, Perry NA, et al. G12/13 is activated by acute tethered agonist exposure in the adhesion GPCR ADGRL3. *Nat Chem Biol*. 2020;16(12):1343–1350. https://doi.org/10.1038/s41589-020-0617-7.

24. Jiang BC, Zhang J, Wu B, et al. G protein-coupled receptor GPR151 is involved in trigeminal neuropathic pain through the induction of Gbetagamma/extracellular signal-regulated kinase-mediated neuroinflammation in the trigeminal ganglion. *Pain*. 2021;162(5):1434–1448. https://doi.org/10.1097/j.pain.0000000000002156.

25. Tang H, Shu C, Chen H, Zhang X, Zang Z, Deng C. Constitutively active BRS3 is a genuinely orphan GPCR in placental mammals. *PLoS Biol*. 2019;17(3):e3000175. https://doi.org/10.1371/journal.pbio.3000175.

26. Iguchi T, Sakata K, Yoshizaki K, Tago K, Mizuno N, Itoh H. Orphan G protein-coupled receptor GPR56 regulates neural progenitor cell migration via a G alpha 12/13 and Rho pathway. *J Biol Chem*. 2008;283(21):14469–14478. https://doi.org/10.1074/jbc.M708919200.

27. Foster SR, Hauser AS, Vedel L, et al. Discovery of human signaling systems: Pairing peptides to G protein-coupled receptors. *Cell*. 2019;179(4):895–908 e21. https://doi.org/10.1016/j.cell.2019.10.010.

28. Muroi T, Matsushima Y, Kanamori R, Inoue H, Fujii W, Yogo K. GPR62 constitutively activates cAMP signaling but is dispensable for male fertility in mice. *Reproduction*. 2017;154(6):755–764. https://doi.org/10.1530/REP-17-0333.

29. Cheng Z, Garvin D, Paguio A, Stecha P, Wood K, Fan F. Luciferase reporter assay system for deciphering GPCR pathways. *Curr Chem Genomics*. 2010;4:84–91. https://doi.org/10.2174/1875397301004010084.

30. Watkins LR, Orlandi C. In vitro profiling of orphan G protein coupled receptor (GPCR) constitutive activity. *Br J Pharmacol*. 2021;178(15):2963–2975. https://doi.org/10.1111/bph.15468.

31. Flock T, Hauser AS, Lund N, Gloriam DE, Balaji S, Babu MM. Selectivity determinants of GPCR-G-protein binding. *Nature*. 2017;545(7654):317–322. https://doi.org/10.1038/nature22070.

32. Pandy-Szekeres G, Munk C, Tsonkov TM, et al. GPCRdb in 2018: Adding GPCR structure models and ligands. *Nucleic Acids Res*. 2018;46(D1):D440–D446. https://doi.org/10.1093/nar/gkx1109.

33. Conklin BR, Farfel Z, Lustig KD, Julius D, Bourne HR. Substitution of three amino acids switches receptor specificity of Gq alpha to that of Gi alpha. *Nature*. 1993;363(6426):274–276. https://doi.org/10.1038/363274a0.

34. Ballister ER, Rodgers J, Martial F, Lucas RJ. A live cell assay of GPCR coupling allows identification of optogenetic tools for controlling Go and Gi signaling. *BMC Biol*. 2018;16(1):10. https://doi.org/10.1186/s12915-017-0475-2.

35. Inoue A, Raimondi F, Kadji FMN, et al. Illuminating G-protein-coupling selectivity of GPCRs. *Cell*. 2019;177(7):1933–1947 e25. https://doi.org/10.1016/j.cell.2019.04.044.

36. Lu S, Jang W, Inoue A, Lambert NA. Constitutive G protein coupling profiles of understudied orphan GPCRs. *PLoS One*. 2021;16(4):e0247743. https://doi.org/10.1371/journal.pone.0247743.

37. Schihada H, Shekhani R, Schulte G. Quantitative assessment of constitutive G protein-coupled receptor activity with BRET-based G protein biosensors. *Sci Signal*. 2021;14(699). https://doi.org/10.1126/scisignal.abf1653. eabf1653.

38. Okashah N, Wan Q, Ghosh S, et al. Variable G protein determinants of GPCR coupling selectivity. *Proc Natl Acad Sci U S A*. 2019;116(24):12054–12059. https://doi.org/10.1073/pnas.1905993116.

39. Morri M, Sanchez-Romero I, Tichy AM, et al. Optical functionalization of human Class A orphan G-protein-coupled receptors. *Nat Commun*. 2018;9(1):1950. https://doi.org/10.1038/s41467-018-04342-1.

40. Morita N, Umemoto E, Fujita S, et al. GPR31-dependent dendrite protrusion of intestinal CX3CR1(+) cells by bacterial metabolites. *Nature*. 2019;566(7742):110–114. https://doi.org/10.1038/s41586-019-0884-1.

41. Wang J, Simonavicius N, Wu X, et al. Kynurenic acid as a ligand for orphan G protein-coupled receptor GPR35. *J Biol Chem*. 2006;281(31):22021–22028. https://doi.org/10.1074/jbc.M603503200.

42. Garcia V, Gilani A, Shkolnik B, et al. 20-HETE signals through G-protein-coupled receptor GPR75 (Gq) to affect vascular function and trigger hypertension. *Circ Res*. 2017;120(11):1776–1788. https://doi.org/10.1161/CIRCRESAHA.116.310525.

43. Gomes I, Bobeck EN, Margolis EB, et al. Identification of GPR83 as the receptor for the neuroendocrine peptide PEN. *Sci Signal*. 2016;9(425). https://doi.org/10.1126/scisignal.aad0694. ra43.

44. Flak MB, Koenis DS, Sobrino A, et al. GPR101 mediates the pro-resolving actions of RvD5n-3 DPA in arthritis and infections. *J Clin Invest*. 2020;130(1):359–373. https://doi.org/10.1172/JCI131609.

45. Shi F, Shen JK, Chen D, et al. Discovery and SAR of a series of agonists at orphan G protein-coupled receptor 139. *ACS Med Chem Lett*. 2011;2(4):303–306. https://doi.org/10.1021/ml100293q.

46. Nohr AC, Shehata MA, Hauser AS, et al. The orphan G protein-coupled receptor GPR139 is activated by the peptides: Adrenocorticotropic hormone (ACTH), alpha-, and beta-melanocyte stimulating hormone (alpha-MSH, and beta-MSH), and the conserved core motif HFRW. *Neurochem Int*. 2017;102:105–113. https://doi.org/10.1016/j.neuint.2016.11.012.

47. Yosten GL, Kolar GR, Redlinger LJ, Samson WK. Evidence for an interaction between proinsulin C-peptide and GPR146. *J Endocrinol*. 2013;218(2):B1–B8. https://doi.org/10.1530/JOE-13-0203.

48. Khrimian L, Obri A, Ramos-Brossier M, et al. Gpr158 mediates osteocalcin's regulation of cognition. *J Exp Med*. 2017;214(10):2859–2873. https://doi.org/10.1084/jem.20171320.

49. Gomes I, Aryal DK, Wardman JH, et al. GPR171 is a hypothalamic G protein-coupled receptor for BigLEN, a neuropeptide involved in feeding. *Proc Natl Acad Sci U S A*. 2013;110(40):16211–16216. https://doi.org/10.1073/pnas.1312938110.

50. Le Mercier A, Bonnavion R, Yu W, et al. GPR182 is an endothelium-specific atypical chemokine receptor that maintains hematopoietic stem cell homeostasis. *Proc Natl Acad Sci U S A*. 2021;118(17). https://doi.org/10.1073/pnas.2021596118.

51. de Lau W, Barker N, Low TY, et al. Lgr5 homologues associate with Wnt receptors and mediate R-spondin signalling. *Nature*. 2011;476(7360):293–297. https://doi.org/10.1038/nature10337.

52. Carmon KS, Gong X, Lin Q, Thomas A, Liu Q. R-spondins function as ligands of the orphan receptors LGR4 and LGR5 to regulate Wnt/beta-catenin signaling. *Proc Natl Acad Sci U S A*. 2011;108(28):11452–11457. https://doi.org/10.1073/pnas.1106083108.

53. Yosten GL, Harada CM, Haddock C, et al. GPR160 de-orphanization reveals critical roles in neuropathic pain in rodents. *J Clin Invest.* 2020;130(5):2587–2592. https://doi.org/10.1172/JCI133270.

54. Gurusamy M, Tischner D, Shao J, et al. G-protein-coupled receptor P2Y10 facilitates chemokine-induced CD4 T cell migration through autocrine/paracrine mediators. *Nat Commun.* 2021;12(1):6798. https://doi.org/10.1038/s41467-021-26882-9.

55. Karhu T, Akiyama K, Vuolteenaho O, et al. Isolation of new ligands for orphan receptor MRGPRX1-hemorphins LVV-H7 and VV-H7. *Peptides.* 2017;96:61–66. https://doi.org/10.1016/j.peptides.2017.08.011.

56. Yu H, Zhao T, Liu S, et al. MRGPRX4 is a bile acid receptor for human cholestatic itch. *Elife.* 2019;8. https://doi.org/10.7554/eLife.48431.

57. Meixiong J, Vasavda C, Snyder SH, Dong X. MRGPRX4 is a G protein-coupled receptor activated by bile acids that may contribute to cholestatic pruritus. *Proc Natl Acad Sci U S A.* 2019;116(21):10525–10530. https://doi.org/10.1073/pnas.1903316116.

58. Schoeder CT, Mahardhika AB, Drabczynska A, Kiec-Kononowicz K, Muller CE. Discovery of tricyclic Xanthines as agonists of the cannabinoid-activated orphan G-protein-coupled receptor GPR18. *ACS Med Chem Lett.* 2020;11(10):2024–2031. https://doi.org/10.1021/acsmedchemlett.0c00208.

59. Dupuis N, Laschet C, Franssen D, et al. Activation of the orphan G protein-coupled receptor GPR27 by surrogate ligands promotes beta-Arrestin 2 recruitment. *Mol Pharmacol.* 2017;91(6):595–608. https://doi.org/10.1124/mol.116.107714.

60. MacKenzie AE, Caltabiano G, Kent TC, et al. The antiallergic mast cell stabilizers lodoxamide and bufrolin as the first high and equipotent agonists of human and rat GPR35. *Mol Pharmacol.* 2014;85(1):91–104. https://doi.org/10.1124/mol.113.089482.

61. Jenkins L, Brea J, Smith NJ, et al. Identification of novel species-selective agonists of the G-protein-coupled receptor GPR35 that promote recruitment of beta-arrestin-2 and activate Galpha13. *Biochem J.* 2010;432(3):451–459. https://doi.org/10.1042/BJ20101287.

62. Hu LA, Tang PM, Eslahi NK, Zhou T, Barbosa J, Liu Q. Identification of surrogate agonists and antagonists for orphan G-protein-coupled receptor GPR139. *J Biomol Screen.* 2009;14(7):789–797. https://doi.org/10.1177/1087057109335744.

63. Zhao X, Stein KR, Chen V, Griffin ME, Hang HC. Chemoproteomics of microbiota metabolites reveals small-molecule agonists for orphan receptor GPRC5A. *bioRxiv.* 2021;12(16):472979. https://doi.org/10.1101/2021.12.16.472979.

64. Offermanns S, Simon MI. G alpha 15 and G alpha 16 couple a wide variety of receptors to phospholipase C. *J Biol Chem.* 1995;270(25):15175–15180. https://doi.org/10.1074/jbc.270.25.15175.

65. Zhu T, Fang LY, Xie X. Development of a universal high-throughput calcium assay for G-protein- coupled receptors with promiscuous G-protein Galpha15/16. *Acta Pharmacol Sin.* 2008;29(4):507–516. https://doi.org/10.1111/j.1745-7254.2008.00775.x.

66. Camarda V, Calo G. Chimeric G proteins in fluorimetric calcium assays: Experience with opioid receptors. *Methods Mol Biol.* 2013;937:293–306. https://doi.org/10.1007/978-1-62703-086-1_18.

67. Liu AM, Ho MK, Wong CS, Chan JH, Pau AH, Wong YH. Galpha(16/z) chimeras efficiently link a wide range of G protein-coupled receptors to calcium mobilization. *J Biomol Screen.* 2003;8(1):39–49. https://doi.org/10.1177/1087057102239665.

68. Vasavda C, Zaccor NW, Scherer PC, Sumner CJ, Snyder SH. Measuring G-protein-coupled receptor signaling via radio-labeled GTP binding. *J Vis Exp.* 2017;(124). https://doi.org/10.3791/55561.

69. Harrison C, Traynor JR. The [35S]GTPgammaS binding assay: Approaches and applications in pharmacology. *Life Sci.* 2003;74(4):489–508. https://doi.org/10.1016/j.lfs. 2003.07.005.

70. Lefkowitz RJ. G protein-coupled receptors. III. New roles for receptor kinases and beta-arrestins in receptor signaling and desensitization. *J Biol Chem.* 1998;273 (30):18677–18680. https://doi.org/10.1074/jbc.273.30.18677.

71. Zhao X, Jones A, Olson KR, et al. A homogeneous enzyme fragment complementation-based beta-arrestin translocation assay for high-throughput screening of G-protein-coupled receptors. *J Biomol Screen.* 2008;13(8):737–747. https:// doi.org/10.1177/1087057108321531.

72. Barnea G, Strapps W, Herrada G, et al. The genetic design of signaling cascades to record receptor activation. *Proc Natl Acad Sci U S A.* 2008;105(1):64–69. https://doi. org/10.1073/pnas.0710487105.

73. Kroeze WK, Sassano MF, Huang XP, et al. PRESTO-tango as an open-source resource for interrogation of the druggable human GPCRome. *Nat Struct Mol Biol.* 2015;22(5):362–369. https://doi.org/10.1038/nsmb.3014.

74. Eishingdrelo H, Cai J, Weissensee P, Sharma P, Tocci MJ, Wright PS. A cell-based protein-protein interaction method using a permuted luciferase reporter. *Curr Chem Genomics.* 2011;5:122–128. https://doi.org/10.2174/1875397301105010122.

75. Oakley RH, Hudson CC, Cruickshank RD, et al. The cellular distribution of fluorescently labeled arrestins provides a robust, sensitive, and universal assay for screening G protein-coupled receptors. *Assay Drug Dev Technol.* 2002;1(1 Pt 1): 21–30. https://doi.org/10.1089/154065802761001275.

76. Dogra S, Sona C, Kumar A, Yadav PN. Tango assay for ligand-induced GPCR-beta-arrestin2 interaction: Application in drug discovery. *Methods Cell Biol.* 2016; 132:233–254. https://doi.org/10.1016/bs.mcb.2015.11.001.

77. Dixon AS, Schwinn MK, Hall MP, et al. NanoLuc complementation reporter optimized for accurate measurement of protein interactions in cells. *ACS Chem Biol.* 2016;11(2):400–408. https://doi.org/10.1021/acschembio.5b00753.

78. Haasen D, Wolff M, Valler MJ, Heilker R. Comparison of G-protein coupled receptor desensitization-related beta-arrestin redistribution using confocal and non-confocal imaging. *Comb Chem High Throughput Screen.* 2006;9(1):37–47. https://doi.org/10. 2174/138620706775213921.

79. Garippa RJ, Hoffman AF, Gradl G, Kirsch A. High-throughput confocal microscopy for beta-arrestin-green fluorescent protein translocation G protein-coupled receptor assays using the Evotec opera. *Methods Enzymol.* 2006;414:99–120. https://doi.org/ 10.1016/S0076-6879(06)14007-0.

80. Bowen WP, Wylie PG. Application of laser-scanning fluorescence microplate cytometry in high content screening. *Assay Drug Dev Technol.* 2006;4(2):209–221. https://doi.org/10.1089/adt.2006.4.209.

81. Eggeling C, Brand L, Ullmann D, Jager S. Highly sensitive fluorescence detection technology currently available for HTS. *Drug Discov Today.* 2003;8(14):632–641. https:// doi.org/10.1016/s1359-6446(03)02752-1.

82. Namkung Y, Le Gouill C, Lukashova V, et al. Monitoring G protein-coupled receptor and beta-arrestin trafficking in live cells using enhanced bystander BRET. *Nat Commun.* 2016;7:12178. https://doi.org/10.1038/ncomms12178.

83. Zhou Y, Meng J, Xu C, Liu J. Multiple GPCR functional assays based on resonance energy transfer sensors. *Front Cell Dev Biol.* 2021;9, 611443. https://doi.org/10.3389/ fcell.2021.611443.

84. Hauge Pedersen M, Pham J, Mancebo H, Inoue A, Asher WB, Javitch JA. A novel luminescence-based beta-arrestin recruitment assay for unmodified receptors. *J Biol Chem.* 2021;296:100503. https://doi.org/10.1016/j.jbc.2021.100503.

85. Foster SR, Brauner-Osborne H. Investigating internalization and intracellular trafficking of Gpcrs: New techniques and real-time experimental approaches. *Handb Exp Pharmacol.* 2018;245:41–61. https://doi.org/10.1007/164_2017_57.

86. Peters MF, Vaillancourt F, Heroux M, Valiquette M, Scott CW. Comparing label-free biosensors for pharmacological screening with cell-based functional assays. *Assay Drug Dev Technol.* 2010;8(2):219–227. https://doi.org/10.1089/adt.2009.0232.

87. Doijen J, Van Loy T, Landuyt B, Luyten W, Schols D, Schoofs L. Advantages and shortcomings of cell-based electrical impedance measurements as a GPCR drug discovery tool. *Biosens Bioelectron.* 2019;137:33–44. https://doi.org/10.1016/j.bios.2019.04.041.

88. Morfa CJ, Bassoni D, Szabo A, et al. A Pharmacochaperone-based high-throughput screening assay for the discovery of chemical probes of orphan receptors. *Assay Drug Dev Technol.* 2018;16(7):384–396. https://doi.org/10.1089/adt.2018.868.

89. Jones EM, Jajoo R, Cancilla D, et al. A scalable, multiplexed assay for decoding GPCR-ligand interactions with RNA sequencing. *Cell Syst.* 2019;8(3):254–260 e6. https://doi.org/10.1016/j.cels.2019.02.009.

90. Wallace ML, Huang KW, Hochbaum D, Hyun M, Radeljic G, Sabatini BL. Anatomical and single-cell transcriptional profiling of the murine habenular complex. *Elife.* 2020;9. https://doi.org/10.7554/eLife.51271.

91. GTEx Consortium. The genotype-tissue expression (GTEx) project. *Nat Genet.* 2013;45(6):580–585. https://doi.org/10.1038/ng.2653.

92. Franzen O, Gan LM, Bjorkegren JLM. PanglaoDB: a web server for exploration of mouse and human single-cell RNA sequencing data. *Database (Oxford).* 2019;2019. https://doi.org/10.1093/database/baz046.

93. Cao Y, Zhu J, Jia P, Zhao Z. scRNASeqDB: A Database for RNA-Seq Based Gene Expression Profiles in Human Single Cells. *Genes (Basel).* 2017;8(12). https://doi.org/10.3390/genes8120368.

94. Haddock CJ, Almeida-Pereira G, Stein LM, et al. Signaling in rat brainstem via Gpr160 is required for the anorexigenic and antidipsogenic actions of cocaine- and amphetamine-regulated transcript peptide. *Am J Physiol Regul Integr Comp Physiol.* 2021;320(3):R236–R249. https://doi.org/10.1152/ajpregu.00096.2020.

95. Kakarala KK, Jamil K. Sequence-structure based phylogeny of GPCR class A rhodopsin receptors. *Mol Phylogenet Evol.* 2014;74:66–96. https://doi.org/10.1016/j.ympev.2014.01.022.

96. Susens U, Hermans-Borgmeyer I, Urny J, Schaller HC. Characterisation and differential expression of two very closely related G-protein-coupled receptors, GPR139 and GPR142, in mouse tissue and during mouse development. *Neuropharmacology.* 2006;50(4):512–520. https://doi.org/10.1016/j.neuropharm.2005.11.003.

97. Isberg V, Andersen KB, Bisig C, Dietz GP, Brauner-Osborne H, Gloriam DE. Computer-aided discovery of aromatic l-alpha-amino acids as agonists of the orphan G protein-coupled receptor GPR139. *J Chem Inf Model.* 2014;54(6):1553–1557. https://doi.org/10.1021/ci500197a.

98. Pi M, Faber P, Ekema G, et al. Identification of a novel extracellular cation-sensing G-protein-coupled receptor. *J Biol Chem.* 2005;280(48):40201–40209. https://doi.org/10.1074/jbc.M505186200.

99. Pi M, Wu Y, Quarles LD. GPRC6A mediates responses to osteocalcin in beta-cells in vitro and pancreas in vivo. *J Bone Miner Res.* 2011;26(7):1680–1683. https://doi.org/10.1002/jbmr.390.

100. Rueda P, Harley E, Lu Y, et al. Murine GPRC6A mediates cellular responses to L-amino acids, but not osteocalcin variants. *PLoS One.* 2016;11(1):e0146846. https://doi.org/10.1371/journal.pone.0146846.

101. Jacobsen SE, Norskov-Lauritsen L, Thomsen AR, et al. Delineation of the GPRC6A receptor signaling pathways using a mammalian cell line stably expressing the receptor. *J Pharmacol Exp Ther.* 2013;347(2):298–309. https://doi.org/10.1124/jpet.113.206276.

102. Dunn HA, Orlandi C, Martemyanov KA. Beyond the ligand: Extracellular and transcellular G protein-coupled receptor complexes in physiology and pharmacology. *Pharmacol Rev.* 2019;71(4):503–519. https://doi.org/10.1124/pr.119.018044.

103. Condomitti G, Wierda KD, Schroeder A, et al. An input-specific orphan receptor GPR158-HSPG interaction organizes hippocampal mossy Fiber-CA3 synapses. *Neuron.* 2018;100(1):201–215 e9. https://doi.org/10.1016/j.neuron.2018.08.038.

104. Orlandi C, Omori Y, Wang Y, et al. Transsynaptic binding of orphan receptor GPR179 to Dystroglycan-Pikachurin complex is essential for the synaptic organization of photoreceptors. *Cell Rep.* 2018;25(1):130–145 e5. https://doi.org/10.1016/j.celrep.2018.08.068.

105. Wright MH, Sieber SA. Chemical proteomics approaches for identifying the cellular targets of natural products. *Nat Prod Rep.* 2016;33(5):681–708. https://doi.org/10.1039/c6np00001k.

106. Piazza I, Beaton N, Bruderer R, et al. A machine learning-based chemoproteomic approach to identify drug targets and binding sites in complex proteomes. *Nat Commun.* 2020;11(1):4200. https://doi.org/10.1038/s41467-020-18071-x.

107. Gamo AM, Gonzalez-Vera JA, Rueda-Zubiaurre A, et al. Chemoproteomic approach to explore the target profile of GPCR ligands: Application to 5-HT1A and 5-HT6 receptors. *Chemistry.* 2016;22(4):1313–1321. https://doi.org/10.1002/chem.201503101.

108. Kufareva I, Ilatovskiy AV, Abagyan R. Pocketome: An encyclopedia of small-molecule binding sites in 4D. *Nucleic Acids Res.* 2012;40(Database issue):D535–D540. https://doi.org/10.1093/nar/gkr825.

109. Ngo T, Ilatovskiy AV, Stewart AG, et al. Orphan receptor ligand discovery by pickpocketing pharmacological neighbors. *Nat Chem Biol.* 2017;13(2):235–242. https://doi.org/10.1038/nchembio.2266.

110. Bender BJ, Gahbauer S, Luttens A, et al. A practical guide to large-scale docking. *Nat Protoc.* 2021;16(10):4799–4832. https://doi.org/10.1038/s41596-021-00597-z.

111. Levit Kaplan A, Strachan RT, Braz JM, et al. Structure-based design of a chemical probe set for the 5-HT5A serotonin receptor. *J Med Chem.* 2022;65(5):4201–4217. https://doi.org/10.1021/acs.jmedchem.1c02031.

112. Patil DN, Singh S, Laboute T, et al. Cryo-EM structure of human GPR158 receptor coupled to the RGS7-Gbeta5 signaling complex. *Science.* 2022;375(6576):86–91. https://doi.org/10.1126/science.abl4732.

113. Jeong E, Kim Y, Jeong J, Cho Y. Structure of the class C orphan GPCR GPR158 in complex with RGS7-Gbeta5. *Nat Commun.* 2021;12(1):6805. https://doi.org/10.1038/s41467-021-27147-1.

114. Lin X, Li M, Wang N, et al. Structural basis of ligand recognition and self-activation of orphan GPR52. *Nature.* 2020;579(7797):152–157. https://doi.org/10.1038/s41586-020-2019-0.

115. Zhou Y, Daver H, Trapkov B, et al. Molecular insights into ligand recognition and G protein coupling of the neuromodulatory orphan receptor GPR139. *Cell Res.* 2022;32(2):210–213. https://doi.org/10.1038/s41422-021-00591-w.

116. Cao C, Kang HJ, Singh I, et al. Structure, function and pharmacology of human itch GPCRs. *Nature.* 2021;600(7887):170–175. https://doi.org/10.1038/s41586-021-04126-6.

117. Yang F, Guo L, Li Y, et al. Structure, function and pharmacology of human itch receptor complexes. *Nature.* 2021;600(7887):164–169. https://doi.org/10.1038/s41586-021-04077-y.

CHAPTER FOUR

Asymmetric activation of class C Gpcrs

Hongnan Liu[a,b], Yanjun Li[a,b], and Yang Gao[a,b,c,*]

[a]Department of Cardiology of Sir Run Run Shaw Hospital, Zhejiang University School of Medicine, Hangzhou, China
[b]Liangzhu Laboratory, Zhejiang University Medical Center, Hangzhou, China
[c]Key Laboratory of Cardiovascular Intervention and Regenerative Medicine of Zhejiang Province, Hangzhou, China
*Corresponding author: e-mail address: gaoyang7@zju.edu.cn

Contents

Abstract

Class C G-protein-coupled receptors (GPCRs) comprise a unique GPCR subfamily with large ligand-binding extracellular domains and function as obligate dimers. The recently resolved cryo-EM structures of full-length GABA$_B$, CaSR, and mGlus have revealed that these receptors are activated in an asymmetric manner, leading to G-protein-coupling on one protomer within the receptor dimer. In this review we discuss the mechanisms of asymmetric activation in class C GPCRs and the unique mode of interaction with the inhibitory Gi protein. Upon activation, the two seven-transmembrane domains (7TMs) of class C GPCRs rearrange to form a conserved asymmetric TM6-TM6 interface. In contrast to class A and B GPCRs, G-protein coupling does not involve the cytoplasmic opening of TM6, but is facilitated through the coordination of intracellular loops. Furthermore, positive and negative allosteric modulators (PAMs and NAMs) adopt distinct conformations to regulate the activity of class C GPCRs. Taken together, these recent findings on the mechanism of asymmetric activation of class C GPCRs highlight a novel mechanism of G protein activation and provide new insights into the design of therapeutics targeting these receptors.

Progress in Molecular Biology and Translational Science, Volume 195
ISSN 1877-1173
https://doi.org/10.1016/bs.pmbts.2022.06.012

77

1. Introduction

G protein-coupled receptors (GPCRs) are the largest family of transmembrane receptors in humans.[1] They play key roles in transducing extracellular signals into intracellular effects.[2] GPCRs can be divided into five classes based on sequence homology: A (rhodopsin-like), B_1 (secretin), B_2 (adhesion), C (glutamate), and F (frizzled/taste)[3] and are responsible for the detection of a vast array of ligands including odorants, hormones, neurotransmitters, chemokines and so on.[1,4] Because of extensive ligand binding capacities and accessible binding sites, GPCRs are considered to be ideal pharmacological targets. Approximately 350 non-olfactory human GPCR members are thought to be druggable and 165 of them are already developed into drug targets.[1,5,6] Currently, 34% of FDA-approved drugs recognize GPCRs as their targets.[5–7] These drugs are used in the treatment of a wide range of diseases including central nervous system diseases, cardiovascular diseases and others.[7]

Class C GPCRs is a small unique GPCR subfamily with 22 members that function as obligate dimers, including eight metabotropic glutamate receptors (mGlu1–8), two gamma-aminobutyric acid type B receptors ($GABA_{B1}$ and $GABA_{B2}$), one calcium-sensing receptor (CaSR) and three taste receptors (TAS1R1–3).[8,9] Among class C GPCRs, mGlus modulate synaptic strength and serve as drug targets for neurological disorders.[10] $GABA_B$ receptors are associated with brain and behavioral diseases, including epilepsy, spasticity, anxiety and neuropathic pain.[11] CaSR regulates calcium homeostasis through its actions in the parathyroid gland and kidneys.[12] 16 drugs that target 8 members of class C GPCRs have been approved by the FDA so far.[1] For example, acamprosate, an antagonist of mGlu5, is clinically used as an anti-neoplastic agent, and cinacalcet, a positive allosteric modulator (PAM) of CaSR, is widely used as a calcimimetic for the treatment of secondary hyperparathyroidism.[1]

Like other classes of GPCRs, class C GPCRs contain a signature seven transmembrane (7TM) domain. However, class C GPCR possesses a large extracellular domain (ECD) responsible for ligand binding[13] (Fig. 1). The N-terminal ECD of class C GPCR is composed of a bi-lobed Venus flytrap domain (VFT) and a cysteine-rich domain (CRD) connecting the VFT to the 7TM domain, with the exception of the $GABA_B$ receptors where CRDs are missing.[9] Dimerization is key to class C GPCR function, among

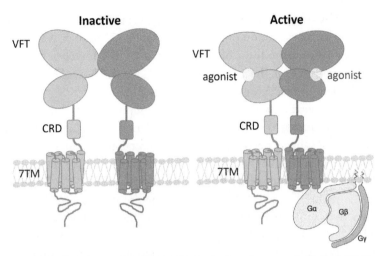

Fig. 1 Cartoon illustration of class C GPCRs. In the inactive state (left), the VFT domains are open holding the CRDs (absent in GABA$_B$) and 7TMs from the two protomers separated. In the active state (right), agonist binding to the one or both protomers lead to the closure of the VFTs bringing the CRDs and 7TMs into close proximity. G protein couples to only one subunit in the dimeric receptor.

which, CaSR form homodimers via disulfide bonds, GABA$_B$ receptors and TAS1Rs form heterodimers, while mGlus can function as either homo- or hetero-dimers.[10,14] Ligand binding at the VFT results in large-scale conformational changes bringing the two 7TMs in close proximity for activation. Activated receptors then turn on various types of cognate G proteins, which modulate a wide range of intracellular effectors.[15] It has been observed that class C GPCR dimers can only bind to one G protein at a time,[16–19] but the coupling mechanism remains elusive. Recent structural studies have begun to reveal the intricate asymmetric mechanisms underlying class C GPCR dimer activation and G protein coupling.

2. Asymmetric activation of class C GPCRs

In class C GCPRs, the activation signal needs to traverse a distance of approximately 120 Å from the agonist-binding site in the VFTs to the G protein-coupled region on the cytoplasmic side of the 7TM. Due to the complexity of the structure and activation mechanism of class C GPCRs, structural studies on class C GPCRs have been challenging. Early X-ray crystallography studies have revealed that agonist binding to ECDs results in the closure and reorientation of the VFTs, and the 7TMs of class C GPCRs are structurally similar to those of class A GPCRs.[20–24] However, how ECD rearrangements result in the activation of the 7TMs and how allosteric modulators regulate the receptor activities remained elusive.

In recent years, with the rapid development of single-particle cryo-electron microscopy (cryo-EM), full-length structures of several class C GPCRs have been resolved, shedding light upon the intricate activation mechanisms of these dimeric GPCRs. Currently, the full-length class C GPCRs resolved include the homodimeric mGlu2,[14,25,26] mGlu4,[25] mGlu5,[13] mGlu7,[14] CaSR[12,27–30] and the heterodimeric $GABA_B$[31–33] in both inactive and active states. In addition, the cryo-EM structures of mGlu2,[25,26] mGlu4[25] and $GABA_B$[33] in complex with the inhibitory Gi protein have also been solved recently. These results provide a fundamental framework for understanding the activation of class C GPCRs. Briefly, agonists binding lead to the closure and rearrangement of the bi-lobed VFTs, bringing the CRDs in close proximity (except for $GABA_B$ receptors that lack CRDs). These conformational changes result in the rearrangement of 7TMs, in which the two protomers exhibit two distinct conformations that enable one of the protomers to couple to G proteins. This activation mechanism, in which the two protomers of homo- and heterodimers exhibit an asymmetric conformation in the active state, is distinctly different from other GPCR classes.

Currently, the high-resolution structural information of class C GPCR heterodimers mainly comes from the $GABA_B$ receptors. $GABA_B$ receptors share conserved VFTs and 7TMs with other class C GPCRs, but do not have the CRD domains. The $GABA_{B1}$ and $GABA_{B2}$ subunits form a obligate heterodimer,[34] in which $GABA_{B1}$ contains an agonist binding site and only $GABA_{B2}$ is capable of coupling to G proteins.[16] $GABA_{B2}$ also allosterically increases the affinity of $GABA_{B1}$ and agonists.[34] In the heterodimeric $GABA_B$-Gi cryo-EM structure, the agonist binds to the VFT of $GABA_{B1}$, resulting in the activation of $GABA_{B2}$ and the coupling of the Gi protein.[33] Upon activation, the 7TM of $GABA_{B1}$ remains unchanged but the $GABA_{B2}$ 7TM undergoes an anti-clockwise rotation relative to $GABA_{B1}$, leading to the formation of an asymmetric TM6–TM6 interface (Fig. 2A). ECL2 of $GABA_B$ has an essential role in relaying structural transitions by ordering the linker that connects the VFT to the 7TM region. Furthermore, comparison between $GABA_{B2}$ structures in the

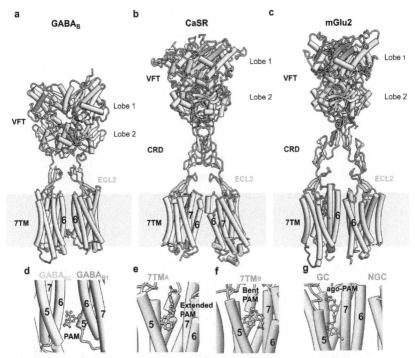

Fig. 2 Asymmetric activation of Class C GPCRs. **a**, Cryo-EM structure of active-state GABAB (PDB: 7EB2). **b**, Cryo-EM structure of active-state CaSR (PDB: 7M3G). **c**, Active-state mGlu2 from the Cryo-EM structure of mGlu2-Gi1 complex (PDB: 7MTS). **d**, Magnified view of active-state GABAB 7TMs complexed with PAM (BHFF). **e**, Magnified view of active-state CaSR 7TM complexed with extended PAM (evocalcet). **f**, Magnified view of active-state CaSR 7TM complexed with bent PAM (evocalcet). **g**, Magnified view of active-state mGlu2 7TM complexed with ago-PAM (ADX55164).

inactive and active states reveals that TM3 and TM5 of $GABA_{B2}$ shift outward upon activation. These movements lead to changes in the ICLs that offer a binding site for the G protein on the intracellular side of the $GABA_{B2}$ subunit. In addition, in active $GABA_B$ dimer, the PAM molecule bound at the bottom of the 7TM interface also acts to stabilize the asymmetric conformation (Fig. 2A, D).

Recently, cryo-EM structures of the CaSR homodimer in inactive and active states were solved by several different research groups.[12,27–30] The higher resolution without enforced C2 symmetry enabled Gao et al. to observe the unique asymmetric configuration in the active-state CaSR

homodimer.[12] Calcium binding leads to closure of the VFTs, bringing the
CRDs together, resulting in the formation of a TM6–TM6 interface at the
7TMs (Fig. 2B). In contrast to the symmetric ECDs, the active-state 7TMs
adopt markedly asymmetric conformations. This asymmetry is also reflected
by the two identical PAMs presenting two different poses in the two protomers (Fig. 2B, E, F). It was demonstrated by inositol monophosphate (IP₁)
accumulation assays with engineered CaSR dimers that the 7TM with an
extended PAM is favored in G-protein coupling.[12] In addition to CaSR,
the structures of full-length homodimeric mGlu2,[25,26] mGlu4,[25] mGlu5[13]
and mGlu7[14] in the active state have also been resolved. Their overall
structural organization is similar to that of CaSR. However, there are some
distinctions between different mGlus. For example, in the inactive state,
the 7TMs of apo mGlu5 are well separated, with TM5 helices being most
proximal between protomers,[13] whereas the 7TMs of inactive mGlu2 come
into proximity to form an asymmetric TM3–TM4 interface.[25,26] The
diverse arrangement of 7TMs in class C GPCRs suggests the possible existence of distinct activation mechanisms and signal transduction modes
among different receptors. In contrast to the diverse 7TM configurations
of different receptors in the inactive state, upon activation, similar
to CaSR, the 7TMs of mGlus rearrange and all form asymmetric
TM6–TM6 interfaces.[13,14,25,26] Similar to CaSR, the asymmetric 7TMs
of mGlu2 are also caused by different interaction networks of ECL2,
ECL3 and CRD in the two protomers (Fig. 2C). Superposition of 7TMs
between the G-protein-coupled (GC) and non-G-protein-coupled
(NGC) subunits of the mGlu2-Gi structure shows conformational differences in TM5 and TM6. An upward shift of TM6 into the extracellular
space occurs in the GC subunit compared to NGC subunit. Another notable
difference between the two subunits is that the agonist-PAM (ago-PAM)
bind the GC subunit in a manner similar to the binding of agonists in
class A GPCRs, whereas it's missing in the NGC subunit (Fig. 2C, G).
The asymmetric binding of ago-PAM further stabilizes the asymmetric
active 7TM configuration. Notably, 7TM bound to ago-PAM exhibits
an open, accessible G-protein-binding pocket on the cytoplasmic side, while
the corresponding pocket is closed in the other protomer. In conclusion,
homodimeric mGlus and CaSR share similar activation patterns, and the
signal of agonist binding to VFTs is transmitted to 7TMs through CRDs,
which further induces asymmetric rearrangement of 7TMs, enabling one
of the 7TMs to couple to G proteins.

Taken together, class C GPCRs share similar asymmetric activation mechanisms enabling one 7TM in the dimer to activate cognate G proteins. In addition, other regulatory molecules such as PAMs also play a role in stabilizing and enhancing the activation potency. In contrast to agonist binding to the conserved VFT, PAMs (or NAMs) binding sites are variable in Class C GPCRs (Fig. 2D–G). Therefore, these binding sites of allosteric modulators may offer more desirable properties in terms of specificity in drug developments compared with the orthosteric ligand-binding sites.

3. G-protein-coupling mechanism in class C GPCRs

The hallmark of G-protein-coupling in class A and B GPCRs is the displacement of the cytosolic part of the TM6 helix to accommodate the insertion of the $G\alpha$ $\alpha5$ helix in a cleft formed at the core of the 7TM.[35,36] The lack of TM6 outward movement observed in the mGlu2-Gi, mGlu4-Gi and GABA$_B$-Gi structures underlines a fundamentally different binding mechanism for class C GPCRs. The structures of mGlu2-G$_{i1}$, mGlu4-G$_{i3}$ and GABA$_B$-G$_{i1}$ complexes exhibit a similar GPCR-Gi binding interface[25,26,33] (Fig. 3A, B). Instead of occupying the deep binding cavity within the transmembrane helical bundle as observed in class A GPCRs (Fig. 3C, D), the C-terminal region of the $G\alpha i$ subunit that has a key role in recognizing GPCRs binds to a shallow groove shaped by ICL1, ICL2 and ICL3 and the intracellular regions of TM3 and TM4 (Fig. 3A–D). ICL2 play a major role in the recognition of Gi (Fig. 3A, B). These data suggest that class C GPCRs share a conserved pattern of G protein recognition distinct from class A GPCRs.

Comparison between the two asymmetric 7TM protomers reveals that the critical elements for G-protein-coupling adopt different conformations. In both Gi-bound mGlu2 and mGlu4 structures, the upwards shift in TM6 of the GC subunit enables the rearrangement of ICL3 hydrophobic residues, which further coordinates the conformations of ICL2, the C terminus and the cytoplasmic tip of TM3, and the engagement with the G protein[25,26]; whereas the structure of active GABA$_B$ reveals an outward shift of intracellular ends of TM3 and TM5 in GABA$_{B2}$ upon activation, resulting in structural relocation of the three ICLs and formation of a shallow groove to accommodate the G protein.[33] In summary, the recent high-resolution structures of class C GPCRs complexed with Gi proteins

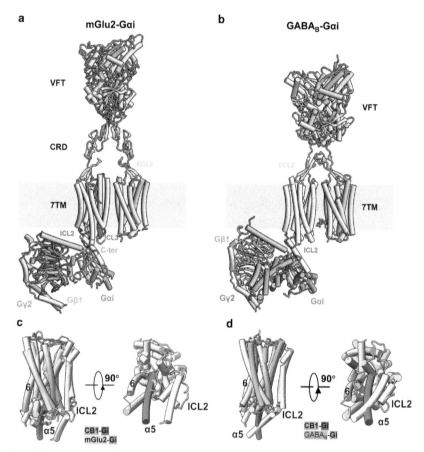

Fig. 3 G-protein-coupling mechanism in class C GPCRs. **a,** Cryo-EM structure of mGlu2-Gi complex (PDB: 7MTS). **b,** Cryo-EM structure of GABA$_B$-Gi complex (PDB: 7EB2). **c,** Comparison of the Gi binding pocket between the mGlu2 and class-A cannabinoid receptor 1 (CB1) (PDB: 6N4B). Two structures were aligned by the class-A TMD as a reference. Colors for α5 of Gαi are: mGlu2-bound, yellow; CB1-bound, red. **d,** Comparison of the Gi binding pocket between the GABA$_B$ and class-A CB1 (PDB: 6N4B). Two structures were aligned by the class-A TMD as a reference. Colors for α5 of Gαi are: GABA$_B$-bound, yellow; CB1-bound, red.

reveal a unique coupling mode different from that observed for monomeric GPCRs (Fig. 3C, D) and this distinct mode of G-protein-coupling, at least with the inhibitory Gi protein, is conserved in class C GPCRs.

4. Prospects

Class C GPCRs are important therapeutic targets and recent cryo-EM studies have begun to reveal the molecular underpinning of the asymmetric activation mechanisms of class C GPCRs and provided new insights for future drug development efforts. In particular, in the class C GPCR dimers, the asymmetric binding of PAMs in the two protomers, as well as the diversity of PAMs binding sites in different receptors, provide new clues and binding-sites for the development of drugs with potentially higher specificity and less side effects. Furthermore, the recently resolved receptor-Gi protein complex structures reveal that class C GPCRs have a unique GPCR-G protein interface distinct from other GPCR classes, where the Gi heterotrimer interact with a shallow groove that involves the ICLs on one of the protomers in the dimer. However, due to the lack of complex structures with different types of G proteins, it is unclear whether this unique coupling mode is conserved in the interactions between class C GPCRs and other G protein families. Therefore, more high-resolution structures of class C GPCRs complexed with various types of different G proteins will be needed to fully unveil the unique G-protein coupling mechanisms of class C GPCR dimers. This will further advance the rational design and discovery of improved therapeutics targeting this physiologically important GPCR subfamily and facilitate the treatment of related diseases.

References

1. Yang D, Zhou Q, Labroska V, et al. G protein-coupled receptors: structure- and function-based drug discovery. *Signal Transduct Target Ther.* Jan 8 2021;6(1):7.
2. Congreve M, de Graaf C, Swain NA, Tate CG. Impact of GPCR structures on drug discovery. *Cell.* Apr 2 2020;181(1):81–91.
3. Kobilka BK. G protein coupled receptor structure and activation. *Biochim Biophys Acta.* Apr 2007;1768(4):794–807.
4. Insel PA, Sriram K, Gorr MW, et al. GPCRomics: an approach to discover GPCR drug targets. *Trends Pharmacol Sci.* 2019;40(6):378–387.
5. Hauser AS, Attwood MM, Rask-Andersen M, Schiöth HB, Gloriam DE. Trends in GPCR drug discovery: new agents, targets and indications. *Nat Rev Drug Discov.* 2017;16(12):829–842.
6. Hauser AS, Chavali S, Masuho I, et al. Pharmacogenomics of GPCR drug targets. *Cell.* 2018;172(1). 41–54.e19.
7. Shimada I, Ueda T, Kofuku Y, Eddy MT, Wüthrich K. GPCR drug discovery: integrating solution NMR data with crystal and cryo-EM structures. *Nat Rev Drug Discov.* 2019;18(1):59–82.
8. Møller TC, Moreno-Delgado D, Pin J-P, Kniazeff J. Class C G protein-coupled receptors: reviving old couples with new partners. *Biophys Rep.* 2017;3(4):57–63.
9. Goudet C, Binet V, Prezeau L, Pin J-P. Allosteric modulators of class-C G-protein-coupled receptors open new possibilities for therapeutic application. *Drug Discov Today Ther Strateg.* 2004;1(1):125–133.
10. Levitz J, Habrian C, Bharill S, Fu Z, Vafabakhsh R, Isacoff E. Mechanism of assembly and cooperativity of homomeric and heteromeric metabotropic glutamate receptors. *Neuron.* 2016;92(1):143–159.
11. Frangaj A, Fan QR. Structural biology of GABAB receptor. *Neuropharmacology.* Jul 1 2018;136(pt A):68–79.

12. Gao Y, Robertson MJ, Rahman SN, et al. Asymmetric activation of the calcium-sensing receptor homodimer. *Nature*. Jul 2021;595(7867):455–459.
13. Koehl A, Hu H, Feng D, et al. Structural insights into the activation of metabotropic glutamate receptors. *Nature*. Feb 2019;566(7742):79–84.
14. Du J, Wang D, Fan H, et al. Structures of human mGlu2 and mGlu7 homo- and heterodimers. *Nature*. Jun 2021;594(7864):589–593.
15. Messa PAC, Brezzi B. Cinacalcet: pharmacological and clinical aspects. *Expert Opin Drug Metab Toxicol*. 2008;4(12):1551–1560.
16. Galvez T, Duthey B, Kniazeff J, et al. Allosteric interactions between GB1 and GB2 subunits are required for optimal GABAB receptor function. *EMBO J*. 2001;20 (9):2152–2159. https://doi.org/10.1093/emboj/20.9.2152.
17. Hlavackova V, Goudet C, Kniazeff J, et al. Evidence for a single heptahelical domain being turned on upon activation of a dimeric GPCR. *EMBO J*. Feb 9 2005;24 (3):499–509.
18. Jacobsen SE, Gether U, Bräuner-Osborne H. Investigating the molecular mechanism of positive and negative allosteric modulators in the calcium-sensing receptor dimer. *Sci Rep*. 2017;7(1):46355.
19. Liu J, Zhang Z, Moreno-Delgado D, et al. Allosteric control of an asymmetric transduction in a G protein-coupled receptor heterodimer. *eLife*. 2017;6:e26985.
20. Geng Y, Mosyak L, Kurinov I, et al. Structural mechanism of ligand activation in human calcium-sensing receptor. *Elife*. 2016;5:e13662.
21. Zhang C, Zhang T, Zou J, et al. Structural basis for regulation of human calcium-sensing receptor by magnesium ions and an unexpected tryptophan derivative co-agonist. *Sci Adv*. 2016;2(5), e1600241.
22. Geng Y, Bush M, Mosyak L, Wang F, Fan QR. Structural mechanism of ligand activation in human GABA(B) receptor. *Nature*. 2013;504(7479):254–259.
23. Dore AS, Okrasa K, Patel JC, et al. Structure of class C GPCR metabotropic glutamate receptor 5 transmembrane domain. *Nature*. Jul 31 2014;511(7511):557–562.
24. Wu H, Wang C, Gregory KJ, et al. Structure of a class C GPCR metabotropic glutamate receptor 1 bound to an allosteric modulator. *Science*. Apr 4 2014;344(6179):58–64.
25. Lin S, Han S, Cai X, et al. Structures of Gi-bound metabotropic glutamate receptors mGlu2 and mGlu4. *Nature*. Jun 2021;594(7864):583–588.
26. Seven AB, Barros-Alvarez X, de Lapeyriere M, et al. G-protein activation by a metabotropic glutamate receptor. *Nature*. Jul 2021;595(7867):450–454.
27. Ling S, Shi P, Liu S, et al. Structural mechanism of cooperative activation of the human calcium-sensing receptor by Ca^{2+} ions and L-tryptophan. *Cell Res*. Apr 2021;31(4):383–394.
28. Chen X, Wang L, Cui Q, et al. Structural insights into the activation of human calcium-sensing receptor. *eLife*. 2021;10:e68578.
29. Park J, Zuo H, Frangaj A, et al. Symmetric activation and modulation of the human calcium-sensing receptor. *Proc Natl Acad Sci U S A*. 2021;118(51):e2115849118.
30. Wen T, Wang Z, Chen X, et al. Structural basis for activation and allosteric modulation of full-length calcium-sensing receptor. *Sci Adv*. 2021;7(23):eabg1483.
31. Papasergi-Scott MM, Robertson MJ, Seven AB, Panova O, Mathiesen JM, Skiniotis G. Structures of metabotropic GABAB receptor. *Nature*. Aug 2020;584 (7820):310–314.
32. Shaye H, Ishchenko A, Lam JH, et al. Structural basis of the activation of a metabotropic GABA receptor. *Nature*. Aug 2020;584(7820):298–303.
33. Shen C, Mao C, Xu C, et al. Structural basis of GABAB receptor-Gi protein coupling. *Nature*. Jun 2021;594(7864):594–598.

34. Marshall FH, Jones KA, Kaupmann K, Bettler B. GABAB receptors-the first 7TM heterodimers. *Trends Pharmacol Sci.* 1999;20(10):396–399.
35. Rasmussen SG, DeVree BT, Zou Y, et al. Crystal structure of the beta2 adrenergic receptor-Gs protein complex. *Nature.* 2011;477(7366):549–555.
36. Zhang Y, Sun B, Feng D, et al. Cryo-EM structure of the activated GLP-1 receptor in complex with a G protein. *Nature.* Jun 8 2017;546(7657):248–253.

> CHAPTER FIVE

Common and selective signal transduction mechanisms of GPCRs

Berkay Selçuk[a] and Ogün Adebali[a,b,*]
[a]Molecular Biology, Genetics and Bioengineering Program, Faculty of Engineering and Natural Sciences, Sabanci University, Istanbul, Turkey
[b]TÜBİTAK Research Institute for Fundamental Sciences, Gebze, Turkey
*Corresponding author: e-mail address: oadebali@sabanciuniv.edu

Contents

Abstract

G protein-coupled receptors (GPCRs) are coupled by four major subfamilies of G proteins. GPCR coupling is processed through a combination of common and selective activation mechanisms together. Common mechanisms are shared for a group of receptors. Recently, researchers managed to identify shared activation pathways for the GPCRs belonging to the same subfamilies. On the other hand, selective mechanisms are responsible for the variations within activation mechanisms. Selective processes can regulate subfamily-specific interactions between the receptor and the G proteins, and intermediate receptor conformations are required to couple particular G proteins through G protein-specific activation mechanisms.

Moreover, G proteins can also selectively interact with RGS (regulators of G protein signaling) proteins as well. Selective processes modulate the signaling profile of the receptor and the tissue they are present. This chapter summarizes the recent research conducted on common and selective signal transduction mechanisms within GPCRs from an evolutionary perspective.

Progress in Molecular Biology and Translational Science, Volume 195
ISSN 1877-1173
https://doi.org/10.1016/bs.pmbts.2022.06.030

1. Introduction

G protein-coupled receptors (GPCRs) are essential components of signal transduction for eukaryotic organisms. These receptors are activated by recognizing external stimuli, such as photons, hormones, or peptides. With more than 800 human receptors, GPCRs are the largest group of transmembrane receptors,[1] which suggests that they are useful membrane-bound tools for "inventing" new functions through gene duplication and divergence during evolution. The number of GPCR genes and sequences are well adapted to the species-specific needs.[2–4] GPCR activation is a complex process involving the cooperation of multiple protein families. In the human genome, there are 4 major GPCR classes (A, B, C, and F), and coupling 16 subtypes of G proteins activate these receptors. Moreover, G proteins' activity is also regulated by a group of proteins called regulators of G protein signaling (RGS). Researchers have been developing novel methodologies to characterize molecular mechanisms of receptor activation.

This review divided receptor activation into two significant subcategories as common and selective activation. Common activation mechanisms can be described as common and conserved mechanisms for a class of GPCRs, while selective activation mechanisms are significantly variable for a range of receptors associated with functional divergence.[5,6] Functional divergence occurs when the new copy of a gene shows a functional difference from its ancestral version after gene duplication. Neofunctionalization is observed when a gene acquires novel functions such as binding to a new ligand and/or coupling to a different g protein. Subfunctionalization is observed when the latest copy of the gene retrieves only certain functions from the ancestor. Suppose the ancestral receptor is coupled to both Gi and Gs subfamilies, but the new copy of the receptor can couple to Gi only. In that case, this is considered an example of subfunctionalization. However, gene duplication can sometimes lead to non-functionalization, resulting in pseudogene formation. By detecting differences in receptor sequence and structure, researchers can reveal the underlying basis of common events such as class-specific activation mechanisms and selective events such as G protein coupling, RGS-G protein interaction, and ligand binding. To fully understand the GPCR activation, we should consider selective and common processes together because they complement each other. Here, we summarized recent studies focusing on different stages of GPCR activation, synthesized the status-quo of GPCR activation research, and discussed how common and selective processes might have evolved.

2. Common activation

GPCRs exhibit a wide range of diversity with respect to their ligands, activation mechanisms, and the G proteins they are coupled to. The main reason why GPCRs are evolutionarily successful for inventing new functions is that residues responsible for core receptor functions such as proper folding and class-specific activation mechanisms are conserved. From an evolutionary perspective, each class can be considered an effective solution to the GPCR activation, which guarantees successful G protein engagement. Differences in receptor functions are variations built on a class-specific foundation they have. Researchers have revealed common mechanisms for individual classes of receptors with the recent increase in the number of experimental GPCR structures.

In 2019 Zhou et al. identified a common activation mechanism shared within class A GPCRs[7] by analyzing 234 experimentally determined structures. This study proposed a new measure to quantify contact between two residues called residue-residue contact score (RRCS). This algorithm was used to determine commonly observed changes in contact score (ΔRRCS) of class A GPCRs upon receptor activation. These common increases or decreases in contact scores were used to build the common activation pathway connecting the bottom of the ligand-binding pocket to G protein coupling. The identified molecular pathway involved important motifs such as CWxP, DRY, Na + pocket, NPxxY, and PIF[8] and integrated them into a single molecular path. The common mechanism was divided into four different layers reflecting the sequential nature of the events occurring within the receptor in extracellular to intracellular direction. These layers did not include any ligand contact because the residues within the ligand-binding pocket are highly variable and well-adapted to the ligand they bind to. Although it has been previously shown that TM6 tilt was observed as the common switch for receptor activation in class A, the series of structural rearrangements leading to that hadn't been described before in detail. Lastly, site-directed mutagenesis of hotspots resulted in constitutionally active or inactive receptors with high accuracy. This suggests that the proposed pathway includes critical hotspots and contacts that can regulate receptor activation.

In 2021 Hauser et al. identified class-specific activation pathways for classes A, B1, C, and F.[9] Their study classified active or inactive state stabilizers interactions based on the differences in contact frequencies in active and inactive-state receptor structures. If an interaction between two residue pairs is observed more frequently in active state structures, it is called an active-state stabilizer and vice versa. Because the number of available structures is imbalanced between classes, they have used different thresholds to determine state stabilizer interactions for each class instead of looking for consensus of every class. In our opinion, this approach is likely to yield false positive determinants rather than false negatives. For example, although they identified 17 ligand contacting state determinants in total, this number is likely to be decreased if the thresholds and the number of available structures are increased, especially for the classes B1, C, and F. However, because their ligands are not highly diverged as in class A, the likelihood of sharing a common ligand contact in these classes should be higher. The methodology for identifying state-stabilizing contacts was implemented into a useful web interface in GPCRdb for public use: https://gpcrdb.org/structure_comparison/comparative_analysis.

Comparison of interactions across classes reveals that most of the contacts for classes A, C, and F were inactivating, while only 39% of the contacts were labeled as inactivating for class B1. The reason why class B1 evolved to have fewer inactivating contacts can be an evolutionary adaptation to lower the energy barrier for achieving the most significant outward TM6 movement observed in any other class. This idea should be supported with further research. Additional to the contact analysis, the analysis of helical movements for each class shows that the cytosolic movement of TM6 is a universal feature for GPCR activation. However, contrary to other classes, no extracellular movement of TM6 was observed within class C, which disproves that TM6 tilt is a common indicator of activation for all GPCRs. Similar to the previously proposed common activation pathway for class A,[7] switches were less tolerant to the alanine mutations than randomly chosen positions. Classes A and B1 shared four state switching residues. In contrast, no other two classes shared any switches suggesting that these classes are evolutionarily closer to each other when compared with other classes (Fig. 1A).

Interestingly, no state determining switches were revealed within the activation mechanism of class C. This can be due to the activation mechanism in class C GPCRs requiring dimerization[10] of the receptors rather than achieving large helical rearrangements. In this case, successful dimerization

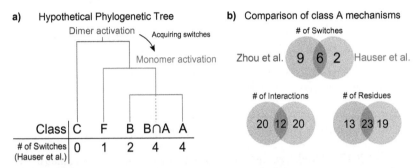

Fig. 1 Comparison of state-determining switches and Class A determinants. (A) A hypothetical phylogenetic tree describing the possible transition between dimer-state to monomer-state activation of GPCRs. (B) Comparison of switches, interactions, and residues across different studies for class A activation mechanisms.

and the interactions within the dimerization interface can act as an activation switch. On the other hand, because receptors in classes A, B1, and F can successfully be activated without any requirement for dimerization, the transition from dimer-state activation to monomer-state activation might have necessitated the invention of new state-determining switches that are triggered upon binding of an agonist to the receptors. In parallel to this hypothesis, the single switch residue of class F is located at the extracellular portion of the receptor responsible for sensing the Wnt ligand.[11] Fig. 1A shows a hypothetical phylogenetic tree built on the existing and shared switches between classes which may not reflect the genuine phylogenetic relationship between classes. Although class C has been shown to be the oldest class previously,[12,13] the relationship between other classes should be further established through a maximum-likelihood-based phylogenetic tree-based approach.

To summarize, researchers identified mechanistic similarities observed during activation for the GPCRs belonging to the same classes with the recent increase in the number of available experimental structures. Comparison of the class A mechanism proposed by these two studies shows that 12 interactions (one with opposite sign), 23 nodes, and 6 switches are shared between two studies (Fig. 1B). Differences between the two networks reflect the differences in the methodology. Both studies emphasized the importance of TM3 in maintaining receptor stability and being a hub for receptor activation, while TM4 made the least number of contacts. TM3 should be necessary for maintaining a functional 7TM fold even for ancestral GPCRs as well due to its central location. In the future, methodologies based on the strength of residue contacts should also be applied to classes B, C, and F as well. While receptors have conserved class-specific features based on the class they belonged to, there are also significant differences between the receptors within each class. These differences are mainly responsible for receptor-specific selective events such as ligand binding and G protein coupling, which we will explore in the following section.

3. Selective processes in G protein coupling

In 2017, Flock et al. performed an evolutionary analysis on 16 human G proteins and identified selectivity determining positions within the known G protein-coupling interface. It has been shown that receptors evolve much faster than the G proteins, and the number of GPCRs is much more excessive than G proteins. Therefore, GPCRs must have evolved features to recognize the selectivity determining barcodes residue composition of G proteins within the coupling interface, to be compatible with the G proteins they are coupled to. Key and lock analogy is being used to describe the selective nature of the G protein-GPCR interactions (Fig. 2). As each key (GPCRs) contains a specific combination of ridges that determines which locks (G proteins) it can open, the coupling ability of each GPCR is determined based on the ridges they evolved for G protein recognition. Moreover, no universal solution has been identified to recognize these G protein-specific barcodes. Thus, each receptor or receptor subfamily should be investigated individually to identify residual determinants of G protein coupling selectivity.

To identify positions determining coupling selectivity, it is crucial to have well-established coupling profiles of receptors. Previously, Guide to the Pharmacology database[14] was the primary resource containing results of the once performed profiling experiments. However, because the experiments had been conducted with different methodologies, the results of these experiments were not consistent with each other in terms of the quantified magnitude of the receptor activity. To overcome this problem, two recent studies,[15,16] revealed coupling profiles of more than 100 human receptors that are also widely used as drug targets. In 2019, Inoue et al. used the idea that coupling selectivity is mainly achieved through the interactions between the C-terminus of the G protein and the receptor,[17] as suggested in the key and lock analogy. By swapping sequences at the C-terminal tail of the G_q 11 unique chimeric G proteins were produced to represent all 16 human G proteins. The main disadvantage of this methodology is its insensitivity to the positions determining selectivity at the backbone of the G protein.

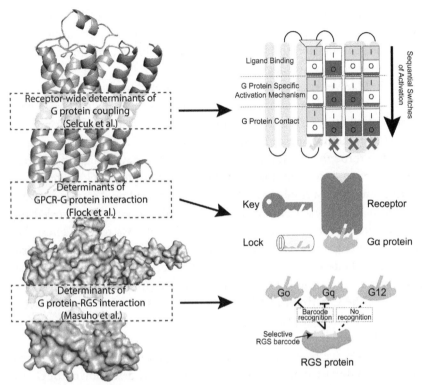

Fig. 2 Three main models describing three selective events. Selcuk et al., Sequential switches of activation model; Flock et al., Key and lock model; Masuho et al., Selectivity barcodes on RGS proteins.

Furthermore, using G_q as a backbone might favor or disfavor specific receptors having a high affinity for G_q. Coupling profiles for 148 human receptors were revealed by measuring the activation of these chimeric G proteins. No motif or conservation patterns were associated with the receptor coupled to similar G proteins. This supports the idea that there is no universal solution for determining selectivity. A logistic regression-based machine learning classifier was produced and shared for public use to identify highly informative features for the receptors with similar coupling profiles https://github.com/raimondifranc/gpcr_coupling_predictor. The important discriminative features were mainly distributed to the coupling interface and contact network of the receptor, suggesting that the coupling interface does not solely determine coupling selectivity. Still, the intermediate receptor conformations could be significant. Using the most discriminative features of their logistic regression-based predictor, researchers have successfully constructed two $G_{12/13}$-coupled DREADDs (designer receptors exclusively activated by designer drugs) by combining ICL3 regions of either GPR183 or GPR132 with the M_3 muscarinic receptor. This has validated the importance of ICL3 in determining G protein coupling selectivity, as suggested before.[18]

In 2022, Avet et al. came up with an improved methodology that does not require modification of the G protein at all.[19] Their biosensors were mainly based on measuring association or disassociation of effector proteins such as p63-RhoGEF, Rap1GAP, and PDZ-RhoGEF to quantify receptor activation through a BRET-based approach. They have managed to profile 100 receptors based on their capability to activate 12 G proteins and recruit three βarrestin subtypes. They have identified new coupling neither shown in Inoue's dataset nor Gtopdb.[14] Furthermore, they revealed that 81% of the receptors were coupled to G15 and 73% to Gz. Thus, they have suggested that a G15/Gz biosensor can be a helpful tool for drug screenings. Differences and similarities between datasets of Inoue et al. and Avet et al. can guide us to identify selectivity determining positions outside of the C-terminus. When these two datasets are analyzed in detail, it could be possible to predict the effect of selectivity determining positions within the coupling interface and the rest of the receptor.

Therefore, we have analyzed the coupling information from these two precious datasets to identify potential receptor-wide determinants of G protein coupling selectivity shared across all aminergic receptors. Aminergic receptors were analyzed through extensive phylogenetic analyses because there is no universal solution for achieving G protein selectivity. It was advantageous to study aminergic receptors because they are one of the mainly investigated GPCRs. As a result, a list of possible selectivity determining positions for the G proteins covering 11 different subtypes was analyzed. These residues were quite dispersed to the receptor's various parts, including ligand-binding pocket, core region, and G protein coupling interface. For a residue to determine coupling selectivity, it should be part of G protein-specific mechanisms. Accordingly, different pathways were identified for G_s, G_{i1}, G_o, and G_q by comparing the contact changes upon coupling to a G protein. These networks were built using only the contact changes between conserved residue pairs and validated the involvement of potential selectivity determining positions in G protein-specific activation mechanisms. The G_s and G_q coupled receptors network included residues from the ligand-binding pocket and extracellular loops, suggesting that receptors favor specific G proteins based on the ligands they are bound to. Although receptors are often coupled to G_{i1} and G_o pathways, different activation mechanisms identified for these receptors suggest that these mechanistic differences can create a basis for selectivity between these two G proteins. The model proposed by selectivity determinants is complementary and inclusive to the findings of Flock et al.; it verifies the role of existing interface residues in coupling selectivity and further suggests the existence of the part of non-interface amino acids covering the entire receptor. Processes controlling coupling selectivity were summarized within a model called "sequential switches of activation" (Fig. 2), which includes three main switches sequentially controlling G protein coupling selectivity: ligand binding, G protein selective activation mechanisms, and G protein contact. According to this model, every switch should be turned on sequentially to achieve successful G protein coupling. Alterations in these three main switches can affect coupling selectivity and increase/decrease the receptor activity.

Besides the selectivity between receptors and G proteins, the interaction between RGS and G proteins is also selective. There are 20 canonical members of RGS proteins in mammals. These proteins regulate the duration of G protein activation by promoting GTPase activity which results in inactivation of the induced signaling pathway.[20] In 2020, Mashuo et al. investigated the sources of G protein and RGS protein selectivity in an extensive study.[21] G protein profiles of all canonical human RGS proteins were revealed through a BRET-based methodology. It was shown that while some RGS subfamilies regulate a wide range of G proteins, others are specialized for certain G protein subtypes.

Moreover, it was demonstrated that no RGS protein could regulate G_s, G_{olf}, G_{12}, and G_{13}. These G proteins have likely evolved not to engage with any canonical RGSs. Certain amino acid substitutions can reverse this situation. Still, these substitutions are likely to be disease-causing because they will alter the signaling profile of a healthy tissue.[21] Similar to the key and lock analogy proposed previously, the RGS and G protein interface was investigated as a source for selectivity (Fig. 2).

The analysis of evolutionarily conserved residues yielded 14 highly conserved positions shared among all RGS proteins and 17 specifically conserved residues present at the periphery of the 14 consensus residues. These 17 residues have been denoted as the significant source for selectivity because they can indirectly regulate RGS and G protein interaction. It was shown that constructing chimeric RGS proteins through swapping their selectivity barcodes can switch their selectivity profiles successfully. On the other hand, partial barcodes were not enough to change the G protein profiles of the RGS proteins. Ancestral reconstruction of extinct RGS proteins revealed that earlier RGS versions were modulating only a few G proteins in a more balanced way in terms of the magnitude of the inhibition. This suggests that they have become more specialized to regulate certain G protein subfamilies during evolution and gained the ability to regulate more G protein subtypes.

4. Mutational characterization of activity determining residues

In 2020, Jones et al. published an extensive study revealing the impact of all possible amino acid substitutions of ADRB2 on the Gs activation.[22] This study reported G_s activation scores for 7828 possible missense variants at four ligand concentrations. It was shown that TM regions and helix 8 are more sensitive to mutations than the loop regions. Moreover, the mutational tolerance of residues was highly correlated (Spearman's $\rho = -0.74$) with their conservation of other orthologous receptors. Among the 15 most intolerant residues, 11 belonged to motifs observed in the common activation mechanism of class A,[7,9] water-mediated interactions, orthosteric ligand-binding pocket. We have characterized[23] the role of second-most intolerant residue 7×41 in receptor activation through molecular dynamics simulations. Significantly, the analysis of molecular dynamics simulations suggested that glycine at 7×41 has a crucial role in G_s-coupled aminergic receptors for achieving a larger TM6 tilt. Deep mutagenesis data shows that non-coupler variants in aminergic receptors predominantly decrease G_s coupling. However, this downregulation can be due to a change in the general activation mechanism of the receptor as well. This dataset should be further analyzed and enriched with different approaches to gain more insight into selective and common receptor functions.

5. Conclusion

Since the first crystal structure rhodopsin[24] was determined, researchers determined more than 600 receptor structures.[25] With the recent cryo-EM revolution, this number is likely to increase further. Various research groups have analyzed existing structures and revealed class-specific activation mechanisms for all human GPCR classes.[7,9] Furthermore, thanks to recent advancements in methodologies to profile receptors and RGS proteins in terms of their G protein preferences, we now have multiple high-quality datasets[15,16,21] that can guide future studies. When these datasets are enriched through phylogeny-based analyses,[13,23] they provide broad insights into selective signal transduction. We expect similar studies for different subfamilies of receptors and uninvestigated GPCR-interacting proteins such as receptor activity-modifying proteins (RAMPs).

Competing interests

The authors declare that they have no conflict of interest.

References

1. Fredriksson R, Lagerström MC, Lundin L-G, Schiöth HB. The G-protein-coupled receptors in the human genome form five main families. Phylogenetic analysis, paralogon groups, and fingerprints. *Mol Pharmacol.* 2003;63(6):1256–1272.
2. Insel PA, Snead A, Murray F, et al. GPCR expression in tissues and cells: Are the optimal receptors being used as drug targets? *Br J Pharmacol.* 2012;165(6):1613–1616. https://doi.org/10.1111/j.1476-5381.2011.01434.x.
3. Niimura Y. Evolutionary dynamics of olfactory receptor genes in chordates: Interaction between environments and genomic contents. *Hum Genomics.* 2009;4(2):107. https://doi.org/10.1186/1479-7364-4-2-107.
4. Schiöth HB, Nordström KJ, Fredriksson R. Mining the gene repertoire and ESTs for G protein-coupled receptors with evolutionary perspective. *Acta Physiol (Oxf).* 2007;190 (1):21–31. https://doi.org/10.1111/j.1365-201X.2007.01694.x.
5. Fay JC, Wu C-I. Sequence divergence, functional constraint, and selection in protein evolution. *Annu Rev Genomics Hum Genet.* 2003;4(1):213–235. https://doi.org/10.1146/annurev.genom.4.020303.162528.
6. Gu X. Functional divergence in protein (family) sequence evolution. *Genetica.* 2003;118 (2–3):133–141.
7. Zhou Q, Yang D, Wu M, et al. Common activation mechanism of class a GPCRs. *Elife.* 2019;8. https://doi.org/10.7554/elife.50279.
8. Filipek S. Molecular switches in GPCRs. *Curr Opin Struct Biol.* 2019;55:114–120. https://doi.org/10.1016/j.sbi.2019.03.017.
9. Hauser AS, Kooistra AJ, Munk C, et al. GPCR activation mechanisms across classes and macro/microscales. *Nat Struct Mol Biol.* 2021;28(11):879–888. https://doi.org/10.1038/s41594-021-00674-7.
10. Zhang XC, Liu J, Jiang D. Why is dimerization essential for class-C GPCR function? New insights from mGluR1 crystal structure analysis. *Protein Cell.* 2014;5(7):492–495. https://doi.org/10.1007/s13238-014-0062-z.
11. Voloshanenko O, Gmach P, Winter J, Kranz D, Boutros M. Mapping of Wnt-frizzled interactions by multiplex CRISPR targeting of receptor gene families. *FASEB J.* 2017;31(11):4832–4844. https://doi.org/10.1096/fj.201700144R.
12. Cvicek V, Goddard III WA, Abrol R. Structure-based sequence alignment of the transmembrane domains of all human GPCRs: Phylogenetic, structural and functional implications. *PLoS Comput Biol.* 2016;12(3):e1004805. https://doi.org/10.1371/journal.pcbi.1004805.
13. Flock T, Hauser AS, Lund N, Gloriam DE, Balaji S, Babu MM. Selectivity determinants of GPCR-G-protein binding. *Nature.* 2017;545(7654):317–322. https://doi.org/10.1038/nature22070.

14. Harding SD, Armstrong JF, Faccenda E, et al. The IUPHAR/BPS guide to PHARMACOLOGY in 2022: Curating pharmacology for COVID-19, malaria and antibacterials. *Nucleic Acids Res.* 2021. https://doi.org/10.1093/nar/gkab1010.

15. Avet C, Mancini A, Breton B, et al. Effector membrane translocation biosensors reveal G protein and βarrestin coupling profiles of 100 therapeutically relevant GPCRs. *bioRxiv.* 2022. https://doi.org/10.1101/2020.04.20.052027.

16. Inoue A, Raimondi F, Kadji FMN, et al. Illuminating G-protein-coupling selectivity of GPCRs. *Cell.* 2019;177(7):1933–1947. https://doi.org/10.1016/j.cell.2019.04.044.

17. Okashah N, Wan Q, Ghosh S, et al. Variable G protein determinants of GPCR coupling selectivity. *Proc Natl Acad Sci.* 2019. https://doi.org/10.1073/pnas.1905993116.

18. Wess J. Molecular basis of receptor/G-protein-coupling selectivity. *Pharmacol Ther.* 1998;80(3):231–264.

19. Avet C, Mancini A, Breton B, et al. Effector membrane translocation biosensors reveal G protein and βarrestin coupling profiles of 100 therapeutically relevant GPCRs. *Elife.* 2022;11:e74101. https://doi.org/10.7554/eLife.74101.

20. Stewart A, Fisher RA. Introduction: G protein-coupled receptors and RGS proteins. *Prog Mol Biol Transl Sci.* 2015;133:1–11. https://doi.org/10.1016/bs.pmbts.2015.03.002.

21. Masuho I, Balaji S, Muntean BS, et al. A global map of G protein signaling regulation by RGS proteins. *Cell.* 2020;183(2):503–521. https://doi.org/10.1016/j.cell.2020.08.052.

22. Jones EM, Lubock NB, Venkatakrishnan A, et al. Structural and functional characterization of G protein–coupled receptors with deep mutational scanning. *eLife.* 2020;9. https://doi.org/10.7554/elife.54895.

23. Selçuk B, Erol I, Durdağı S, Adebali O. Evolutionary association of receptor-wide amino acids with G protein coupling selectivity in aminergic GPCRs. *bioRxiv.* 2022. https://doi.org/10.1101/2021.09.15.460528.

24. Palczewski K, Kumasaka T, Hori T, et al. Crystal structure of rhodopsin: A G protein-coupled receptor. *Science.* 2000;289(5480):739–745. https://doi.org/10.1126/science.289.5480.739.

25. Kooistra AJ, Mordalski S, Pándy-Szekeres G, et al. GPCRdb in 2021: Integrating GPCR sequence, structure and function. *Nucleic Acids Res.* 2021;49(D1):D335–D343. https://doi.org/10.1093/nar/gkaa1080.

CHAPTER SIX

Appreciating the potential for GPCR crosstalk with ion channels

Amy Davies and Alejandra Tomas*

Section of Cell Biology and Functional Genomics, Department of Metabolism, Digestion and Reproduction, Imperial College London, London, United Kingdom
*Corresponding author: e-mail address: a.tomas-catala@imperial.ac.uk

Contents

Abstract

G protein-coupled receptors (GPCRs) are expressed by most tissues in the body and are exploited pharmacologically in a variety of pathological conditions including diabetes, cardiovascular disease, neurological diseases, and cancers. Numerous cell signaling pathways can be regulated by GPCR activation, depending on the specific GPCR, ligand and cell type. Ion channels are among the many effector proteins downstream of these signaling pathways. Saliently, ion channels are also recognized as druggable targets, and there is evidence that their activity may regulate GPCR function via membrane potential and cytoplasmic ion concentration. Overall, there appears to be a large potential for crosstalk between ion channels and GPCRs. This might have implications not only for targeting GPCRs for drug development, but also opens the possibility of co-targeting them with ion channels to achieve improved therapeutic outcomes. In this review, we highlight the large variety of possible GPCR-ion channel crosstalk modes.

Progress in Molecular Biology and Translational Science, Volume 195
ISSN 1877-1173
https://doi.org/10.1016/bs.pmbts.2022.06.013

101

Abbreviations

AC	adenylate cyclase
AKAP	A kinase anchoring protein
AR	adrenergic receptor
cAMP	cyclic AMP
CaV	voltage gated calcium channels
DAG	diacylglycerol
DR	dopamine receptor
FRET	fluorescent resonance energy transfer
GIPR	gastric inhibitory peptide receptor
GIRK	G protein-coupled inwardly rectifying potassium channel
GLP-1R	glucagon like peptide 1 receptor
GluR	glutamate receptor
GPCR	G protein coupled receptor
IBP	intracellular binding partners
IP3	inositol triphosphate
MOR	mu opioid receptor
MR	muscarinic receptor
PDE	phosphodiesterase
PIP2	phosphatidylinositol 4,5-bisphosphate
PKA	protein kinase A
PKC	protein kinase C
PLC	phospholipase C
TIRF	total internal reflection fluorescence
TRP	transient receptor potential channels
TSHR	thyroid stimulating hormone receptor

1. Introduction

Transmembrane (TM) proteins span the width of the membrane and as such have residues exposed extracellularly and intracellularly. GPCRs are important TM proteins, being widely expressed in the human body and representing a large group of druggable protein targets for the treatment of conditions such as neurological and metabolic diseases. This is due to their ability to robustly up- or down-regulate key signaling pathways in cells with functional consequences, and their accessibility from the blood plasma. GPCR activation and signaling culminates in a change in behavior in various effector proteins and thus, a change in cell function. Examples of current GPCR drug therapies include Lixisenatide which binds to the glucagon-like peptide-1 receptor (GLP-1R) and is used for type 2 diabetes (T2D) treatment, as well as Suvorexant which binds to the oxytocin receptors (1/2) to treat insomnia.[1] Notably, there are still many GPCRs with uncharacterized ligands called orphan GPCRs which may be unrecognized drug targets. Furthermore, GPCRs characterized under non-pathological conditions may be underappreciated as targets in diseased states which can alter their function.[2,3] Another group of TM proteins which are recognized to hold pharmaceutical potential are ion channels. Ion channels open and close to regulate the movement of ions between the intracellular and extracellular environment, and may be specific to an ion, or to a particular charge. Ion channels have critical functions such as maintaining the resting membrane potential of animal cells and in the generation of action potentials in electrically excitable cells such as cardiac, neuronal and muscle cells. They also partake in calcium signaling. Although they have not been exploited to the same extent as GPCRs, there are very successful drugs which target ion channels on the market. Examples include amlodipine which is a calcium channel blocker used to treat coronary artery disease, and sulfonylureas, which are K ATP channel openers used to stimulate pancreatic beta cell insulin secretion in T2D.[4]

Importantly, ion channels are common effectors downstream of GPCR signaling, and there is evidence that the activity of ion channels may impact GPCR signaling. Moreover, studies suggest that their communication may not only be dependent on the specific GPCR and the ion channel compliment of a cell, but also on other cell-specific qualities, and on the specific GPCR ligand employed. These findings have implications for the future of GPCR-based therapies in disease. As such, the aim of this review is to highlight the variety in possible GPCR-ion channel bi-directional interactions, and how this may be impacted by other factors.

2. GPCR activation

GPCRs are divided into classes (A, B, C, D, E and F) based on their sequence homology. However, despite a lack of sequence homology between classes, they share a common secondary structure. GPCRs have an extracellular N-terminus, followed by seven TM helices, and an intracellular C-terminus which mediates interactions between the receptor and intracellular binding partners (IBPs). GPCRs are flexible and dynamic, existing in a state of functional equilibrium. Importantly, ligand binding increases the probability of the receptor taking on an active-like conformation which can interact with IBPs.[5,6] IBPs include signaling G proteins and the scaffolding protein beta arrestin; each of these have different functions downstream of their activation.

G proteins are heterotrimeric proteins, consisting of a Gγ, a Gβ and a Gα subunit. In the inactive state, all three subunits are bound to each other, and Gα is bound to a GDP molecule. This heterotrimeric complex can bind to GPCRs. Upon GPCR activation, the conformational change in the GPCR allows the interacting Gα subunit to exchange its bound GDP for a GTP, resulting in its dissociation from the other G protein subunits and the receptor. Both the resulting Gα subunit and the Gβγ heterodimer can induce downstream signaling by interacting with other proteins localized at the lipid membrane to which they are anchored. This signaling capacity remains until regulator of G protein signaling proteins hydrolyses the GTP bound to the Gα subunit, resulting in heterotrimer reassociation. G protein signaling can also be terminated at the level of the GPCR, as a certain binding orientation of the beta arrestin can sterically hinder G protein binding.[7,8]

3. GPCR signaling and ion channel regulation

Ion channels can be regulated by second messengers which are downstream of GPCR activation. Additionally, there is a membrane-delimited method of ion channel regulation which involves the direct interaction of active G protein subunits with the ion channel.

3.1 G protein regulation of ion channels

G proteins, following their release from the heterotrimeric G protein complex by GPCR activation, can directly modulate transmembrane ion channel function. There are examples of ion channels which can bind to Gα as well as Gβγ proteins. It has long been known that Gβγ subunit binding can regulate the activity of calcium voltage gated channels (CaV). Functionally, the impact of Gβγ subunit binding is variable depending on the CaV channel subtype, but for Cav2.1 and Cav2.2, Gβγ binding shifts the voltage-dependence of voltage gating to more depolarized potentials.[9,10] Another channel which is modulated by Gβγ subunits is the TRPM3 cation channel. TRPM3 is effectively inhibited by the direct application of Gβγ subunits in excised inside-out patches,[11,12] and by activation of GPCRs coupled to $Gα_q$, $Gα_s$ and $Gα_i$ in a manner dependent on Gβγ.[12,13] The TRPM1 channel is another example. This channel can interact with both a Gα and a Gβγ subunit, such that they work synergistically together.[14]

There is potential for ion channels to be promiscuously regulated in this way by differently coupled GPCRs, for instance, if they interact with Gβγ or with Gα subunits non-specifically. However, there is documentation of Gα subtype specificity among ion channels. For example, the GIRK channel is activated by Gβγ binding following specific activation of $G\alpha_{i/o}$-coupled receptors. Structural information supports that this might be attributed to $G\alpha_{i/o}$ interaction with the GIRK channel, facilitating Gβγ binding.[15] Gα protein subtype selectivity is also seen with the TRPM8 cation channel and with the TRPC4 channel. Studies suggest that TRPM8 is more potently inhibited by the $G\alpha_q$ subunit than by $G\alpha_{11}$,[16,17] whereas the TRPC4 channel does not directly bind to $G\alpha_q$ but is directly activated by $G\alpha_i$.[18,19] Meanwhile, neither TRPC4 nor TRPM8 appear to directly bind to Gβγ. Overall, this represents a way in which GPCR regulation of ion channels may achieve specificity.

3.2 Second messenger regulation of ion channels

GPCRs may regulate the activity of ion channels via G protein modulation of signaling pathways. $G\alpha_s$ and $G\alpha_i$ proteins are associated with the cyclic AMP second messenger cascade, while $G\alpha_q$ is associated with the PLC signaling pathway. $G\alpha_s$ and $G\alpha_i$ proteins modulate the cAMP pathway by respectively upregulating and downregulating adenylate cyclase (AC) enzymes; these catalyze cAMP synthesis. In turn, cAMP binds to and activates its two major effectors, PKA and Epac. Signaling via the PLC pathway occurs when $G\alpha_q$ proteins activate membrane bound PLC-beta isoforms; these subsequently catalyze the conversion of PIP_2 into the two second messengers IP_3 and DAG, where DAG in turn upregulates PKC. Notably, the Gβ protein subunit can also partake in signaling pathway regulation, with reported roles in modulating the activated PLC beta isoform and various AC isoforms.[20,21] Both protein kinases and second messengers are implicated in regulating the activity of ion channels.

PKC and PKA phosphorylation of ion channels can modulate their activity (Fig. 1). Phosphorylation can have various regulatory effects,

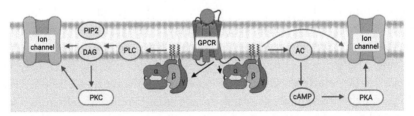

Fig. 1 GPCRs can regulate ion channel function via activating G proteins. This can either be membrane delimited (green arrow) or it can be via second messenger cascades (red arrows). *Created in BioRender.com.*

including ion channel voltage sensitivity, current, as well as their surface expression.[22–26] For example, the ligand-gated chloride channel, GlyR, is internalized in response to PKA and PKC activators.[22] In contrast, in the heterologous xenopus oocyte model, PKC activation increases plasma membrane localization and current through smooth muscle and cardiac Cav1.2 channel isoforms.[26] Recently, it was found that phosphorylation of a specific residue on a variant of a sodium voltage-gated channel increases voltage sensitivity of the channel[23]; this phosphorylation was mediated by PKC but not by PKA, outlining a non-redundant role between the kinases. Importantly, protein kinase phosphorylation can have different effects on the same ion channel if the channel has multiple phosphorylation sites depending on the specific residue which has been phosphorylated. For example, the Kir6.2 subunit of K ATP channels is phosphorylated at Ser372 and at T180. While PKC phosphorylation of Ser372 leads to down-regulation of the channel complex, phosphorylation of the T180 subunit increases the open probability of the channel.[27,28] Similarly, mutating certain combinations of phosphorylation sites on Cav2.3 can uncover either a stimulatory or an inhibitory role for PKC.[29] Supporting physiological relevance for having multiple phosphorylation sites with distinct functional outcomes, different PKC isoforms have varied ability to activate Cav2.2 depending on their relative efficacies at phosphorylating certain combinations of residues.[30] Of additional note is that kinases may indirectly modulate ion channel activity by phosphorylating interacting proteins. For example, PKA can increase Cav1.2 activity by phosphorylating a protein called Rad which otherwise binds to tonically inhibit the channel.[31] In addition, PKC phosphorylation of $\alpha2\delta$-1 results in altered trafficking of the NMDA receptor and ion channel (NMDAR).[32] Therefore, phosphorylation of ion channels is a highly versatile regulatory mechanism with potential to be engaged by active GPCRs.

Interestingly, lipid binding is also implicated in the regulation of ion channel function (Fig. 1). For example, the cryo-EM structure of the human TRPC5 ion channel has revealed a DAG binding site near the pore region of the channel,[33] while experimental studies have demonstrated that the murine TRPC5 is activated by DAG.[34,35] Interestingly, the role of DAG appears to be to facilitate enhanced interactions of the murine TRPC5 with PIP_2, which subsequently activates the channel, rather than being directly responsible for channel opening.[34] In fact, experimental and structural evidence suggests that direct PIP2 modulation is a widespread phenomenon among ion channels.[36] For example, membrane PIP_2 depletion by various methods can inhibit the current amplitude of calcium channels including Cav2.2 and Cav2.1,[37,38] while mutagenesis experiments have verified that a PIP_2 interaction motif identified on the CaV channel is responsible for a large amount of the CaV channels' PIP_2 sensitivity.[39] Additionally, cryo-EM analysis of GIRK channels suggests that PIP_2 interaction permits binding of the Gßy subunit, and thus the activation of the ion channel—consistent with experimental data reporting that PIP_2 is required for GIRK activation[40]; while x-ray crystallography has provided evidence that PIP_2 interaction alters the conformation of the classical inward rectifier K+ channel Kir2.2 to open the channel.[41] Like GIRK channels, Kir2.2 requires PIP_2 for its activity.[42] On the contrary, Kir6.1 appears to be insensitive to PIP_2, despite having a PIP_2 binding site.[43] Overall, current evidence indicates that lipids downstream of G_qPCR signaling can bind to some ion channels and act either as allosteric modulators or direct activators; however, a binding site may not necessarily translate to functionality.

Also, to note is that second messenger signaling is not limited to regulating ion channels at the plasma membrane. Epac, PKA, PKC and IP_3 are all implicated in increasing the opening of ion channels located at the endoplasmic reticulum, and other intracellular ion stores depending on the cell type, with the consequence of altering cytoplasmic calcium concentration.[44-48] It has also more recently been observed that cAMP generated following G_sPCR activation can directly potentiate calcium release via IP_3Rs at the ER.[45]

4. GPCR signal compartmentalization

Evidence suggests that GPCR signaling is spatially confined to specific domains of activity in the membrane and cytosol, such that some downstream proteins are concentrated within these domains while others are excluded. This is achieved by a combination of signal propagation restriction and co-localization of pathway proteins. Importantly, it facilitates specific interactions between a given GPCR and a subset of effector proteins.

4.1 Restriction of signal propagation

Current understanding is that restricting signal propagation prevents all proteins which are sensitive to GPCR signaling in a cell from being ubiquitously affected. In this regard, compartmentalization of the H_2O-soluble and diffusible signal cAMP has received particular interest. In this respect, a general inhibitor of phosphodiesterases (PDEs), the enzymes in charge

of breaking down cAMP, potentiates cAMP responses following endogenous G_sPCR activation.[49] Moreover, it has recently been demonstrated,
using FRET-based "nanorulers," that nanometer-sized cAMP domains surrounding a GPCR can be contained in a manner dependent on endogenous
PDE activity.[50] These studies highlight PDEs as important factors for restriction of cAMP signaling from a given GPCR. Moreover, there is evidence
that such PDE-mediated restriction of cAMP has functional consequences
for downstream effector proteins, including ion channels. Knockout of PDE
isoforms facilitates a larger β-adrenergic G_sPCR-mediated change in downstream effectors.[51] Moreover, in cardiac myocytes pharmacological PDE
inhibition enhances the L-Type CaV channel response to β2AR, β1AR
and GluR GPCRs,[49] as well as cAMP-dependent regulation of IK_s potassium ion channel activity.[52] Saliently, there exist various PDE isoforms
which differ in their coupling to specific GPCRs and/or effector proteins
and in their ability to create cAMP sinks,[49,51–55] facilitating heterogeneity
in signal propagation between specific GPCRs and downstream effector
proteins.[49] In relation to this, it is to note that some PDE isoforms are calcium sensitive and so activity of calcium ion channels may enhance PDE-
mediated cAMP compartmentalization, as recently seen in pancreatic
beta cells.[56]

Modelling studies suggest nevertheless that PDE catalytic activity alone is
insufficient to compartmentalize freely diffusing cAMP.[57] However, other
studies indicate that cAMP diffusion is hindered in cells so that it is not free.
In a recent study, it was observed that the catalytic subunit of PKA forms
condensates under endogenous conditions, reducing cAMP diffusion kinetics, a process vital for PDEs to shield a fused cAMP FRET biosensor from
cAMP.[58,59] In addition, another study found that mitochondrial density significantly correlated with cAMP diffusion kinetics.[60] Cell geometry has also
been demonstrated to shape cAMP diffusion and signaling dynamics, for
example, as seen in cilia and elongated neuronal cells.[61–63] Thus, evidence
supports that a combination of PDE catalytic activity and cAMP diffusion
hindrance will contribute to the shielding of proteins, such as ion channels,
from cAMP generated upon GPCR stimulation.

4.2 Protein co-localization

As would be expected with restricted GPCR signaling, studies indicate that
the co-localization of GPCRs and downstream proteins facilitates efficient
signaling to select targets (Fig. 2). On a macromolecular scale, GPCRs may

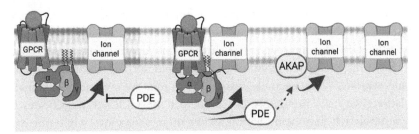

Fig. 2 GPCRs may be coupled to or isolated from ion channels. Signal targeting might be achieved by direct interactions between GPCRs and/or pathway proteins. In some cases, co-localization alone via subcellular segregation of the proteins, such as in lipid rafts (purple membrane), may achieve this. *Created in BioRender.com.*

be physically anchored with downstream signaling and effector proteins. A Kinase Anchoring Proteins (AKAPs) are scaffolding proteins which can bind to various GPCR pathway proteins in an isoform-specific manner including AC, PKA and PKC, and ion channels.[64] An AKAP-organized "signalosome" containing both a GPCR and ion channels has been identified. Specifically, AKAP150/79 can anchor the β-adrenergic GPCR with L–Type calcium channels as well as M–Type K+ channels, TRPV channels, PKA and PKC.[65,66] Remarkably, AKAP150/79 anchoring is vital for β-adrenergic receptor-mediated enhancement of L type calcium voltage gated activity in the brain and heart.[67,68] Even without incorporation of a GPCR, AKAP complexes can still function to amplify GPCR signaling such that an effector protein is modulated at a lower level of GPCR activation than would otherwise be possible.[69] Importantly, complex formation may also facilitate membrane-delimited regulation in some instances. Evidence suggests that both the GIRK channel and GPCRs can associate with the same G protein heterotrimer simultaneously, and that co-localization is key in allowing M2R-mediated ion channel activation.[15,70] Of further interest, recently obtained mass spectrometry data from our laboratory (unpublished) has revealed that the G_s-coupled GPCR GLP-1R is physically associated with the pancreatic K ATP ion channel following its activation. Future research will determine the functional importance of this anchoring, but one may speculate that it might enhance the ability of the GLP-1R to regulate the function of the channel.

In the absence of direct protein–protein interactions, co–localization may still be achieved. It is known that the plasma membrane is highly heterogenous, with preferential lipid mixing resulting in the formation of different lipid-ordered domains. Of note is that in living membranes, protein content will also drive this mixing. As such, the heterogeneity of a cell membrane will depend both on its membrane protein and its lipid content.[71] Saliently, GPCRs, their signaling partners and downstream effectors can be differentially targeted to liquid-ordered (lipid rafts) or -disordered membrane regions.[72] For example, using TIRF microscopy and biochemical isolation, we have previously observed that the GLP-1R translocates to cholesterol-rich membrane nanodomains upon activation, while our evidence suggests that the related GIPR is constitutively localized at lipid rafts.[73] Biochemical isolation has also inferred the raft-localization of numerous ion channels, including various isoforms of potassium, sodium, and calcium voltage gated ion channels, while others are excluded.[74]

Illustrating nicely that co-localization can be important for signal transduction, a recent study found that disrupting PKA-AKAP interaction abolished a FRET signal from a PKA biosensor tethered to an activated GPCR.[50] All in all, it can be inferred that co-localization either via subcellular targeting and/or physical interactions is important for facilitating GPCR regulation of ion channel activity.

5. Voltage and ions modulate GPCR function

The activity of ion channels can change intracellular ion concentration to determine the membrane potential. Therefore, different ion channel compliments confer specific properties of electrical excitability and resting membrane potential to a cell, as well as dynamic cytoplasmic ion concentrations, with evidence supporting that this in turn affects GPCR responses.

5.1 Ions and GPCR regulation

It has been shown for multiple class A GPCRs that sodium can negatively modulate receptor activity.[75] Moreover, high-resolution X-ray crystallography of numerous class A GPCRs has shown that sodium is able to bind to a pocket which is open in the inactive state receptor.[75,76] Consistent with this, mass spectrometry techniques have revealed that agonist binding results in the loss of sodium binding to the GPCR, whereas antagonists retain the ion.[77] Furthermore, mutating various residues in the binding pocket abrogates the sodium-dependent effects on receptor activity (Refs. 78,79; White et al., 2018). Therefore, experimental, and structural studies support that sodium binds this pocket to confer allosteric effects. Providing further insight, modeling studies indicate that sodium is displaced because agonist binding leads to protonation of a key residue—Asp79,[80] whereby the protonated residue can subsequently stabilize an active receptor conformation.[81] Importantly, the sodium binding pocket and the Asp79 residue are highly conserved, and it is predicted that sodium binding is a ubiquitous phenomenon in class A GPCRs.[75] Thus, it is generally accepted that sodium is a common allosteric modulator of this GPCR class. This might have particular importance for cells which have dynamic changes in sodium ion concentration, for example, neuronal cells.

In addition to direct allosteric effects of sodium ions on class A GPCRs, GPCRs may also be regulated by calcium signaling in the cell. For example, an increase in intracellular calcium concentration, either by the mobilization of intracellular calcium stores or influx through plasma membrane channels, can activate the ubiquitous protein kinase calcium calmodulin, which subsequently alters GPCR behavior. Calcium calmodulin binds to six of the seven isoforms of GRKs which are responsible for phosphorylating GPCRs and facilitating beta arrestin binding and internalization. Importantly, early studies demonstrated the ability of calmodulin to inhibit various GRK isoforms, preventing their downregulation of GPCRs.[82,83]

5.2 Evidence for voltage gating of GPCRs

Depolarization has been documented to influence the activity of numerous class A GPCRs, including muscarinic,[84,85] glutamatergic,[86] purinergic,[59] opioid,[87] adrenergic,[88,89] and dopaminergic receptors.[90,91] Moreover, a gating current reminiscent of those in ion channels has been measured, providing evidence that voltage dependence is intrinsic to GPCRs.[84] Specifically, studies report that changes in membrane voltage can affect agonist affinity[84-86,89] and efficacy.[88,89,91] These findings support that voltage can affect both ligand binding and ligand-dependent interactions with intracellular binding partners.

Saliently, findings support that voltage dependency varies greatly depending on the GPCR. For example, using the same experimental model, it was shown that the M2R and the mGluR1 have greater affinity for their endogenous ligands, acetylcholine (Ach) and glutamate, respectively, when the plasma membrane is depolarized, whereas the M1R and the mGluR3 have reduced affinity.[85,86] In addition, dopamine potency at the D3R was unaffected by voltage, whereas this was decreased for D1R and D5R upon membrane depolarization.[90,92] The general mechanism for voltage dependence may also be receptor dependent. For the M2R, PTX treatment eliminated the voltage dependency of the receptor; this is consistent with voltage modulating only high-affinity receptor states in which they are bound to the G protein.[85,93] In contrast, PTX treatment had no apparent effect on the voltage dependency of alpha2A-AR,[89] although in this instance, it is notable that ligand-specific effects have also been documented and may contribute to mechanistic differences (as discussed in Section 5).

Considering the physiological relevance of voltage modulation of GPCRs,[84] Ben-Chaim et al.[84]showed that the ON gating current in M2R-Ach ternary complex at small depolarizations has both a fast and a slow component, suggesting that a single action potential should move the fast-gating charge, whereas the slow-gating charge might require a train of action potentials. This would have implications for how different electrical activity of cells may modulate receptor signaling. In addition, recent studies have provided evidence that voltage modulation of GPCR signaling can have significant impacts on downstream physiological processes. In a recent study, P2YR was mutated to convey voltage insensitivity and it was found that voltage regulation of the receptor is important for modulating the quantal size release of catecholamine downstream of P2YR activation in chromaffin cells.[59] Saliently, this study also demonstrated that catecholamine collected from adrenal slices was only in sufficient concentration to influence cardiac myocyte contractability when the slice was subject to membrane depolarization.[59] Another group successfully used CRISPR technology to generate a fly strain expressing voltage-insensitive drosophila muscarinic receptors.[94] This study showed that voltage modulation of the receptor is important for normal synaptic release and neuronal potentiation, with the mutant receptor-expressing flies exhibiting altered behavior; specifically, they displayed increased odor habituation. There is need to further investigate the importance of GPCR-voltage dependency in the nervous system but also in other electrically excitable cells, such as pancreatic and cardiac cells.[94]

Notably, it has been suggested that the gating charge is a result of the movement of a sodium ion through the GPCR. This suggestion came from modeling of the receptor, where a change in membrane voltage was simulated to alter the position of the ion in the protein.[80,95] Friedman et al.[79] sought to experimentally test this: they found that mutating the charged residues implicated in direct interaction with the sodium ion resulted in reduced agonist affinity of the M2R. This effect parallels that of the voltage-dependent effects on agonist affinity. However, they also demonstrated that the mutation did not attenuate the voltage dependence of M2R agonist affinity, indicating that in fact sodium allosteric modulation and voltage modulation of the receptor are two separate modulatory events.[79]

6. Ligand—specific GPCR activity

Ligands with different structures, within the realm of being able to bind to the GPCR orthosteric site, can confer specific behavioral properties to GPCRs. Structurally, this is because of differences in interaction between the ligand and residues in the site, which are then propagated. Equally, changes in the GPCR orthosteric site can alter a ligand's ability to transduce a signal via the GPCR. These are important considerations when developing new GPCR-based therapies, such that ligands which enhance beneficial GPCR signaling and are effective under physiological conditions are designed or selected for.

GPCRs are often promiscuous: commonly, a particular GPCR can couple to more than one G protein isoform as well as to beta arrestins. Importantly, 'biased' ligands can favorably couple a GPCR to specific intracellular binding partners, and resultantly up–regulate some signaling pathways more than others (Fig. 3A). For example, various biased agonists of the GLP-1R have been documented and characterized by our group. These include the exogenous agonists exendin-F1 and exendin-P5 which are biased toward G proteins compared to the endogenous agonist GLP-1.[96] There can also be bias between G protein subtypes. For example, a small molecule drug can bias the TSH Receptor towards $G\alpha_q$ over $G\alpha_s$ recruitment.[97] Furthermore, not only do biased ligands exhibit relative differences in signal pathway recruitment at the level of the G protein, but they can also be associated with specific receptor trafficking signatures, including, as previously shown by us, differential plasma membrane motilities of individually tracked

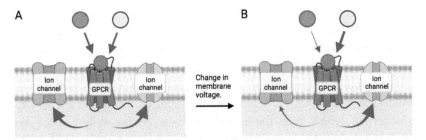

Fig. 3 GPCR—ion channel regulation may be dependent on the ligand and ion channel activity. (A, B) Purple, agonist 1, Green, agonist 2. (A) Ligands conveying certain signaling bias may differently couple a GPCR to downstream ion channels. (B) A change in membrane voltage, caused by a change in ion channel activity, may alter the efficacy and/or affinity of select ligands. *Created in BioRender.com.*

receptors and nanodomain clustering propensities at the cell membrane.[73,98] Thus, signal compartmentalization may also be regulated by bias.[98,99] Ultimately, bias can translate to differences in cell behavior and physiology.[96,97] By modulating bias, it may be possible to target an activated GPCR towards or away from an ion channel, with physiologically relevant effects.

Interestingly, in this context there is also evidence that the voltage dependency of GPCRs is ligand-specific (Fig. 3). This has been documented for the M2R, MOR, D2R and βARs. A change in voltage has been documented to alter ligand-induced receptor activity and/or ligand binding by a magnitude dependent on the ligand.[87,88,91,100–102] A change in voltage may also have opposite effects on ligand-GPCR interactions across a set of ligands. This was for example seen when analyzing MOR activation by 10 different ligands using FRET: for some ligands, membrane depolarization enhanced ligand-induced activation whereas for others, it reduced activation. Interestingly, for one MOR ligand it was even seen that a change in voltage switched its pharmaceutical class between an antagonist and an agonist.[87] For the D2R, it was reported that depolarization may reduce ligand efficacy and/or affinity across a set of D2R ligands[91]; this supports the idea that, for some GPCRs, differences in both the GPCR pre-active and ternary complex determines ligand-specific voltage dependence. Altogether, it is apparent that the ion channel activity in a target cell, which might change upon GPCR activation, needs to be considered when screening GPCR ligands.

7. Conclusions

Evidence suggests that signaling pathways engaged by GPCRs are linked to various aspects of ion channel function, in a manner specific to the ion channel. Evidence also suggests that there may exist a two-way regulatory relationship. Ion channel activity influences plasma membrane resting membrane potential and action potential properties, as well as cytoplasmic ion concentration. These may have functional consequences for some GPCRs more than for others. Importantly, the ability of a given GPCR to couple to an ion channel may be ligand specific. It may also depend on the expression and activity of various molecules other than the ion channel and the GPCR themselves; these include molecules involved in signaling or signal restriction, or molecules mediating the co-localization or physical interaction of the involved proteins. On a larger scale, this process may depend on organelle organization, inter-organelle contacts and geometry of the cell. These observations have consequences for understanding the biology of different cell types, but also for conditions where these factors may change.

Considering therapeutic potential, by elucidating how a given GPCR interacts, or avoids interactions, with given ion channels, it might be possible to design new biased ligands which strengthen, weaken, or even establish novel interactions. Biased agonism is already recognized to hold the potential to increase tissue selectivity; in this context, a GPCR may only couple with a given ion channel via a specific pathway in some cell types. Similarly, if a ligand is more able to transduce a signal at a certain membrane voltage, and cells in the body exhibit distinct plasma membrane potentials, it might be possible to design a ligand which has best affinity and/or efficacy at specifically targeted sites. This would presumably reduce side effects and improve tolerability among the population. GPCR—ion channel interactions could also be important in the movement towards personalized medicine. If an individual has a genomic variation or disease which alters the activity of an ion channel, then this may influence the therapeutic choice of GPCR ligand. This might involve for example selecting a ligand with different voltage sensitivity. Additionally, it could involve redirecting the GPCR away from interaction with said ion channel and focusing on an alternate GPCR—ion channel relationship. Lastly, it might be possible to co-target ion channels directly to enhance a GPCR ligand effectiveness.

Overall, a variety of potential GPCR—ion channel crosstalk exists. These relationships may be exploited pharmacologically in a variety of ways. Therapeutically targeting certain GPCR—ion channel interactions, which may be cell type- and condition-dependent, will involve developing future GPCR ligands which confer specific bias and/or voltage sensitivity.

References

1. Hauser AS, Attwood MM, Rask-Andersen M, Schiöth HB, Gloriam DE. Trends in GPCR drug discovery: new agents, targets and indications. *Nat Rev Drug Discov.* 2017;16(12):829–842.
2. Nikolaev VO, Moshkov A, Lyon AR, et al. Beta2-adrenergic receptor redistribution in heart failure changes cAMP compartmentation. *Science.* 2010;327(5973):1653–1657.
3. Oduori OS, Murao N, Shimomura K, et al. Gs/Gq signaling switch in β cells defines incretin effectiveness in diabetes. *J Clin Investig.* 2020;130(12):6639–6655.
4. Bagal SK, Brown AD, Cox PJ, et al. Ion channels as therapeutic targets: a drug discovery perspective. *Journal of Medicinal Chemistry.* 2013;56(3):593–624.
5. Frei JN, Broadhurst RW, Bostock MJ, et al. Conformational plasticity of ligand-bound and ternary GPCR complexes studied by ^{19}F NMR of the β_1-adrenergic receptor. *Nat Commun.* 2020;11(1):669. https://doi.org/10.1038/s41467-020-14526-3.

6. Solt AS, Bostock MJ, Shrestha B, et al. Insight into partial agonism by observing multiple equilibria for ligand-bound and G_s-mimetic nanobody-bound β_1-adrenergic receptor. *Nat Commun.* 2017;8:1795. https://doi.org/10.1038/s41467-017-02008-y.

7. Nguyen AH, Thomsen A, Cahill 3rd TJ, et al. Structure of an endosomal signaling GPCR-G protein-β-arrestin megacomplex. *Nat Struct Mol Biol.* 2019;26(12): 1123–1131. https://doi.org/10.1038/s41594-019-0330-y.

8. Thomsen A, Plouffe B, Cahill TJ, 3rd, et al. GPCR-G protein-β-arrestin super-complex mediates sustained G protein signaling. *Cell.* 2016;166(4):907–919. https://doi.org/10.1016/j.cell.2016.07.004.

9. Bean BP. Neurotransmitter inhibition of neuronal calcium currents by changes in channel voltage dependence. *Nature.* 1989;340(6229):153–156.

10. Zamponi GW, Currie KP. Regulation of Ca(V)2 calcium channels by G protein coupled receptors. *Biochim Biophys Acta.* 2013;1828(7):1629–1643.

11. Badheka D, Yudin Y, Borbiro I, et al. Inhibition of transient receptor potential melastatin 3 ion channels by G-protein βγ subunits. *eLife.* 2017;6:e26147.

12. Quallo T, Alkhatib O, Gentry C, Andersson DA, Bevan S. G protein βγ subunits inhibit TRPM3 ion channels in sensory neurons. *eLife.* 2017;6:e26138.

13. Alkhatib O, da Costa R, Gentry C, Quallo T, Bevan S, Andersson DA. Promiscuous G-protein-coupled receptor inhibition of transient receptor potential melastatin 3 ion channels by Gβγ subunits. *The Journal of Neuroscience.* 2019;39 (40):7840–7852.

14. Xu Y, Orlandi C, Cao Y, et al. The TRPM1 channel in ON-bipolar cells is gated by both the α and the βγ subunits of the G-protein Go. *Sci Rep.* 2016;6:20940.

15. Kano H, Toyama Y, Imai S, et al. Structural mechanism underlying G protein family-specific regulation of G protein-gated inwardly rectifying potassium channel. *Nat Commun.* 2019;10(1):2008.

16. Li L, Zhang X. Differential inhibition of the TRPM8 ion channel by Gαq and Gα 11. *Channels (Austin, Tex).* 2013;7(2):115–118.

17. Zhang X. Direct Gα$_q$ gating is the sole mechanism for TRPM8 inhibition caused by bradykinin receptor activation. *Cell Rep.* 2019;27(12):3672–3683.e4. https://doi.org/10.1016/j.celrep.2019.05.080.

18. Jeon JP, Hong C, Park EJ, et al. Selective Gαi subunits as novel direct activators of transient receptor potential canonical (TRPC)4 and TRPC5 channels. *J Biol Chem.* 2012;287(21):17029–17039.

19. Myeong J, Kwak M, Jeon JP, Hong C, Jeon JH, So I. Close spatio-association of the transient receptor potential canonical 4 (TRPC4) channel with Gαi in TRPC4 activation process. *Am J Physiol Cell Physiol.* 2015;308(11):C879–C889.

20. Pfeil EM, Brands J, Merten N, et al. Heterotrimeric G protein subunit gαq is a master switch for Gβγ-mediated calcium mobilization by Gi-coupled GPCRs. *Mol Cell.* 2020;80(6):940–954.e6.

21. Smrcka AV. G protein βγ subunits: central mediators of G protein-coupled receptor signaling. *Cell Mol Life Sci.* 2008;65(14):2191–2214.

22. Breitinger U, Bahnassawy LM, Janzen D, et al. PKA and PKC modulators affect ion channel function and internalization of recombinant alpha1 and alpha1-beta glycine receptors. *Front Mol Neurosci.* 2018;11:154.

23. Kerth CM, Hautvast P, Körner J, Lampert A, Meents JE. Phosphorylation of a chronic pain mutation in the voltage-gated sodium channel Nav1.7 increases voltage sensitivity. *J Biol Chem.* 2021;296:100227.

24. Park CG, Park Y, Suh BC. The HOOK region of voltage-gated Ca2+ channel β subunits senses and transmits PIP2 signals to the gate. *J Gen Physiol.* 2017;149(2): 261–276.

25. Qian H, Patriarchi T, Price JL, et al. Phosphorylation of Ser1928 mediates the enhanced activity of the L-type Ca2+ channel Cav1.2 by the β2-adrenergic receptor in neurons. *Sci Signal.* 2017;10(463):eaaf9659.

26. Raifman TK, Kumar P, Haase H, Klussmann E, Dascal N, Weiss S. Protein kinase C enhances plasma membrane expression of cardiac L-type calcium channel, $Ca_V 1.2$. *Channels (Austin, Tex)*. 2017;11(6):604–615.

27. Aziz Q, Thomas AM, Khambra T, Tinker A. Regulation of the ATP-sensitive potassium channel subunit, Kir6.2, by a Ca^{2+}-dependent protein kinase C. *The Journal of Biological Chemistry*. 2012;287(9):6196–6207.

28. Light PE, Bladen C, Winkfein RJ, Walsh MP, French RJ. Molecular basis of protein kinase C-induced activation of ATP-sensitive potassium channels. *Proc Natl Acad Sci USA*. 2000;97(16):9058–9063.

29. Rajagopal S, Burton BK, Fields BL, El IO, Kamatchi GL. Stimulatory and inhibitory effects of PKC isozymes are mediated by serine/threonine PKC sites of the $Ca_v 2.3\alpha_1$ subunits. *Arch Biochem Biophys*. 2017;621:24–30.

30. Rajagopal S, Fang H, Oronce CI, et al. Site-specific regulation of CA(V)2.2 channels by protein kinase C isozymes betaII and epsilon. *Neuroscience*. 2009;159(2):618–628.

31. Katz M, Subramaniam S, Chomsky-Hecht O, et al. Reconstitution of β-adrenergic regulation of $Ca_V 1.2$: Rad-dependent and Rad-independent protein kinase A mechanisms. *Proc Natl Acad Sci USA*. 2021;118(21):e2100021118.

32. Zhou MH, Chen SR, Wang L, et al. Protein kinase C-mediated phosphorylation and α2δ-1 interdependently regulate NMDA receptor trafficking and activity. *J Neurosci*. 2021;41(30):6415–6429.

33. Storch U, Mederos y Schnitzler M, Gudermann T. A greasy business: identification of a diacylglycerol binding site in human TRPC5 channels by cryo-EM. *Cell Calcium*. 2021;97:102414.

34. Ningoo M, Plant LD, Greka A, Logothetis DE. PIP_2 regulation of TRPC5 channel activation and desensitization. *J Biol Chem*. 2021;296:100726.

35. Storch U, Forst AL, Pardatscher F, et al. Dynamic NHERF interaction with TRPC4/5 proteins is required for channel gating by diacylglycerol. *Proc Natl Acad Sci USA*. 2017;114(1):E37–E46.

36. Suh BC, Hille B. PIP2 is a necessary cofactor for ion channel function: how and why? *Annu Rev Biophys*. 2008;37:175–195. https://doi.org/10.1146/annurev.biophys.37. 032807.

37. de la Cruz L, Puente EI, Reyes-Vaca A, et al. PIP_2 in pancreatic β-cells regulates voltage-gated calcium channels by a voltage-independent pathway. *Am J Physiol Cell Physiol*. 2016;311(4):C630–C640.

38. Vivas O, Castro H, Arenas I, Elías-Viñas D, García DE. PIP2 hydrolysis is responsible for voltage independent inhibition of CaV2.2 channels in sympathetic neurons. *Biochem Biophys Res Commun*. 2013;432(2):275–280.

39. Kaur G, Pinggera A, Ortner NJ, et al. A polybasic plasma membrane binding motif in the I–II linker stabilizes voltage-gated $Ca_v 1.2$ calcium channel function. *J Biol Chem*. 2015;290:21086–21100.

40. Niu Y, Tao X, Touhara KK, MacKinnon R. Cryo-EM analysis of PIP_2 regulation in mammalian GIRK channels. *eLife*. 2020;9:e60552.

41. Hansen SB, Tao X, MacKinnon R. Structural basis of PIP2 activation of the classical inward rectifier K+ channel Kir2.2. *Nature*. 2011;477(7365):495–498.

42. D'Avanzo N, Cheng WW, Doyle DA, Nichols CG. Direct and specific activation of human inward rectifier K+ channels by membrane phosphatidylinositol 4,5-bisphosphate. *J Biol Chem*. 2010;285(48):37129–37132.

43. Quinn KV, Cui Y, Giblin JP, Clapp LH, Tinker A. Do anionic phospholipids serve as cofactors or second messengers for the regulation of activity of cloned ATP-sensitive K+ channels? *Circ Res*. 2003;93(7):646–655.

44. Ferris CD, Huganir RL, Bredt DS, Cameron AM, Snyder SH. Inositol trisphosphate receptor: phosphorylation by protein kinase C and calcium calmodulin-dependent protein kinases in reconstituted lipid vesicles. *Proc Natl Acad Sci USA*. 1991;88 (6):2232–2235.

45. Konieczny V, Tovey SC, Mataragka S, Prole DL, Taylor CW. Cyclic AMP recruits a discrete intracellular Ca^{2+} store by unmasking hypersensitive IP_3 receptors. *Cell Rep.* 2017;18(3):711–722.

46. Pereira L, Cheng H, Lao DH, et al. Epac2 mediates cardiac β1-adrenergic-dependent sarcoplasmic reticulum Ca2+ leak and arrhythmia. *Circulation.* 2013;127(8):913–922.

47. Pereira L, Métrich M, Fernández-Velasco M, et al. The cAMP binding protein Epac modulates Ca^{2+} sparks by a Ca^{2+}/calmodulin kinase signalling pathway in rat cardiac myocytes. *J Physiol.* 2007;583(pt 2):685–694.

48. Wojcikiewicz RJ, Luo SG. Phosphorylation of inositol 1,4,5-trisphosphate receptors by cAMP-dependent protein kinase. Type I, II, and III receptors are differentially susceptible to phosphorylation and are phosphorylated in intact cells. *J Biol Chem.* 1998;273(10):5670–5677.

49. Rochais F, Abi-Gerges A, Horner K, et al. A specific pattern of phosphodiesterases controls the cAMP signals generated by different Gs-coupled receptors in adult rat ventricular myocytes. *Circ Res.* 2006;98(8):1081–1088.

50. Anton SE, Kayser C, Maiellaro I, et al. Receptor-associated independent cAMP nanodomains mediate spatiotemporal specificity of GPCR signaling. *Cell.* 2022;185 (7):1130–1142.e11.

51. Blackman BE, Horner K, Heidmann J, et al. PDE4D and PDE4B function in distinct subcellular compartments in mouse embryonic fibroblasts. *The Journal of Biological Chemistry.* 2011;286:12590–12601.

52. Terrenoire C, Houslay MD, Baillie GS, Kass RS. The cardiac IKs potassium channel macromolecular complex includes the phosphodiesterase PDE4D3. *J Biol Chem.* 2009;284:9140–9146.

53. Bock A, Annibale P, Konrad C, et al. Optical mapping of cAMP signaling at the nanometer scale. *Cell.* 2020;182(6):1519–1530.e17.

54. Nikolaev VO, Bünemann M, Schmitteckert E, Lohse MJ, Engelhardt S. Cyclic AMP imaging in adult cardiac myocytes reveals far-reaching beta1-adrenergic but locally confined beta2-adrenergic receptor-mediated signaling. *Circ Res.* 2006;99(10): 1084–1091.

55. Surdo N, Berrera M, Koschinski A, et al. FRET biosensor uncovers cAMP nano-domains at β-adrenergic targets that dictate precise tuning of cardiac contractility. *Nat Commun.* 2017;8:15031.

56. Tenner B, Getz M, Ross B, et al. Spatially compartmentalized phase regulation of a Ca^{2+}-cAMP-PKA oscillatory circuit. *eLife.* 2020;9:e55013.

57. Yang PC, Boras BW, Jeng MT, et al. A computational modeling and simulation approach to investigate mechanisms of subcellular cAMP compartmentation. *PLoS Comput Biol.* 2016;12(7):e1005005.

58. Zaccolo M, Zerio A, Lobo MJ. Subcellular organization of the cAMP signaling pathway. *Pharmacol Rev.* 2021;73(1):278–309.

59. Zhang Q, Liu B, Li Y, Zhou Z. Regulating quantal size of neurotransmitter release through a GPCR voltage sensor. *Proc Natl Acad Sci USA.* 2020;117(43):26985–26995.

60. Richards M, Lomas O, Jalink K, et al. Intracellular tortuosity underlies slow cAMP diffusion in adult ventricular myocytes. *Cardiovasc Res.* 2016;110(3):395–407.

61. Li L, Gervasi N, Girault JA. Dendritic geometry shapes neuronal cAMP signalling to the nucleus. *Nat Commun.* 2015;6:6319.

62. Truong ME, Bilekova S, Choksi SP, et al. Vertebrate cells differentially interpret ciliary and extraciliary cAMP. *Cell.* 2021;184(11):2911–2926.e18.

63. Wu CT, Hilgendorf KI, Bevacqua RJ, et al. Discovery of ciliary G protein-coupled receptors regulating pancreatic islet insulin and glucagon secretion. *Genes Dev.* 2021;35(17-18):1243–1255.

64. Zhang J, Bal M, Bierbower S, Zaika O, Shapiro MS. AKAP79/150 signal complexes in G-protein modulation of neuronal ion channels. *J Neurosci.* 2011;31:7199–7211.

65. Btesh J, Fischer M, Stott K, McNaughton PA. Mapping the binding site of TRPV1 on AKAP79: implications for inflammatory hyperalgesia. *J Neurosci.* 2013;33(21): 9184–9193.

66. Zhang J, Shapiro MS. Mechanisms and dynamics of AKAP79/150-orchestrated multi-protein signalling complexes in brain and peripheral nerve. *J Physiol.* 2016;594 (1):31–37.

67. Davare MA, Avdonin V, Hall DD, et al. A beta2 adrenergic receptor signaling complex assembled with the Ca^{2+} channel Cav1.2. *Science.* 2001;293(5527):98–101.

68. Hall DD, Davare MA, Shi M, et al. Critical role of cAMP-dependent protein kinase anchoring to the L-type calcium channel Cav1.2 via A-kinase anchor protein 150 in neurons. *Biochemistry.* 2007;46(6):1635–1646.

69. Bucko PJ, Scott JD. Drugs that regulate local cell signaling: AKAP targeting as a therapeutic option. *Ann Rev Pharmacol Toxicol.* 2021;61:361–379.

70. Tateyama M, Kubo Y. Gi/o-coupled muscarinic receptors co-localize with GIRK channel for efficient channel activation. *PLoS One.* 2018;13:e0204447.

71. Gómez-Llobregat J, Buceta J, Reigada R. Interplay of cytoskeletal activity and lipid phase stability in dynamic protein recruitment and clustering. *Sci Rep.* 2013;3:2608.

72. Villar VA, Cuevas S, Zheng X, Jose PA. Localization and signaling of GPCRs in lipid rafts. *Methods Cell Biol.* 2016;132:3–23.

73. Buenaventura T, Bitsi S, Laughlin WE, et al. Agonist-induced membrane nanodomain clustering drives GLP-1 receptor responses in pancreatic beta cells. *PLoS Biol.* 2019; 17(8):e3000097.

74. Dart C. Lipid microdomains and the regulation of ion channel function. *J Physiol.* 2010;588(pt 17):3169–3178.

75. Katritch V, Fenalti G, Abola EE, Roth BL, Cherezov V, Stevens RC. Allosteric sodium in class A GPCR signaling. *Trends Biochem Sci.* 2014;39(5):233–244.

76. White KL, Eddy MT, Gao ZG, et al. Structural connection between activation microswitch and allosteric sodium site in GPCR signaling. *Structure.* 2018;26(2):259–269. e5. https://doi.org/10.1016/j.str.2017.12.013.

77. Agasid MT, Sørensen L, Urner LH, Yan J, Robinson CV. The effects of sodium ions on ligand binding and conformational states of G protein-coupled receptors-insights from mass spectrometry. *Journal of the American Chemical Society.* 2021;143(11): 4085–4089.

78. Massink A, Gutiérrez-de-Terán H, Lenselink EB, et al. Sodium ion binding pocket mutations and adenosine A2A receptor function. *Mol Pharmacol.* 2015;87(2):305–313.

79. Friedman S, Tauber M, Ben-Chaim Y. Sodium ions allosterically modulate the M2 muscarinic receptor. *Sci Rep.* 2020;10:11177.

80. Vickery ON, Carvalheda CA, Zaidi SA, Pisliakov AV, Katritch V, Zachariae U. Intracellular transfer of Na^{+} in an active-state g-protein-coupled receptor. *Struct (London, England: 1993).* 2018;26(1):171–180.e2.

81. Fleetwood O, Matricon P, Carlsson J, Delemotte L. Energy landscapes reveal agonist control of G protein-coupled receptor activation via microswitches. *Biochemistry.* 2020;59(7):880–891.

82. Gurevich EV, Tesmer JJ, Mushegian A, Gurevich VV. G protein-coupled receptor kinases: more than just kinases and not only for GPCRs. *Pharmacol Ther.* 2012;133 (1):40–69.

83. Pronin AN, Satpaev DK, Slepak VZ, Benovic JL. Regulation of G protein-coupled receptor kinases by calmodulin and localization of the calmodulin binding domain. *J Biol Chem.* 1997;272(29):18273–18280.

84. Ben-Chaim Y, Chanda B, Dascal N, Bezanilla F, Parnas I, Parnas H. Movement of 'gating charge' is coupled to ligand binding in a G-protein-coupled receptor. *Nature.* 2006;444(7115):106–109.

85. Ben-Chaim Y, Tour O, Dascal N, Parnas I, Parnas H. The M2 muscarinic G-protein-coupled receptor is voltage-sensitive. *The Journal of Biological Chemistry.* 2003;278(25):22482–22491.

86. Ohana L, Barchad O, Parnas I, Parnas H. The metabotropic glutamate G-protein-coupled receptors mGluR3 and mGluR1a are voltage-sensitive. *J Biol Chem.* 2006;281(34):24204–24215.

87. Kirchhofer SB, Lim VJY, Ruland J, Kolb P, Bunemann M. Differential recognition of opioid analgesics by opioid receptors: Predicted interaction patterns correlate with ligand-specific voltage sensitivity. *bioRxiv.* 2021. https://doi.org/10.1101/2021.12.02.470941.

88. Birk A, Rinne A, Bünemann M. Membrane potential controls the efficacy of catecholamine-induced β1-adrenoceptor activity. *The Journal of Biological Chemistry.* 2015;290(45):27311–27320.

89. Rinne A, Birk A, Bünemann M. Voltage regulates adrenergic receptor function. *Proc Natl Acad Sci USA.* 2013;110(4):1536–1541.

90. Ågren R, Sahlholm K. Voltage-dependent dopamine potency at D_1-like dopamine receptors. *Frontiers in Pharmacology.* 2020;11:581151.

91. Sahlholm K, Barchad-Avitzur O, Marcellino D, et al. Agonist-specific voltage sensitivity at the dopamine D2S receptor—molecular determinants and relevance to therapeutic ligands. *Neuropharmacology.* 2011;61(5-6):937–949.

92. Sahlholm K, Marcellino D, Nilsson J, Fuxe K, Arhem P. Differential voltage-sensitivity of D2-like dopamine receptors. *Biochem Biophys Res Commun.* 2008;374(3):496–501. https://doi.org/10.1016/j.bbrc.2008.07.052.

93. Mahaut-Smith MP, Martinez-Pinna J, Gurung IS. A role for membrane potential in regulating GPCRs? *Trends Pharmacol Sci.* 2008;29(8):421–429.

94. Rozenfeld E, Tauber M, Ben-Chaim Y, Parnas M. GPCR voltage dependence controls neuronal plasticity and behavior. *Nat Commun.* 2021;12(1):7252.

95. Vickery ON, Machtens J-P, Tamburrino G, Seeliger D, Zachariae U. Structural mechanisms of voltage sensing in G protein-coupled receptors. *Structure.* 2016;24:997–1007.

96. Marzook A, Tomas A, Jones B. The interplay of glucagon-like peptide-1 receptor trafficking and signalling in pancreatic beta cells. *Front Endocrinol.* 2021;12:678055.

97. Latif R, Morshed SA, Ma R, Tokat B, Mezei M, Davies TF. A Gq biased small molecule active at the TSH receptor. *Front Endocrinol.* 2020;11:372.

98. Fletcher MM, Halls ML, Zhao P, et al. Glucagon-like peptide-1 receptor internalisation controls spatiotemporal signalling mediated by biased agonists. *Biochem Pharmacol.* 2018;156:406–419.

99. Manchanda Y, Bitsi S, Kang Y, Jones B, Tomas A. Spatiotemporal control of GLP-1 receptor activity. *Curr Opin Endocrine Metab Res.* 2021;16:19–27.

100. Moreno-Galindo EG, Alamilla J, Sanchez-Chapula JA, Tristani-Firouzi M, Navarro-Polanco RA. The agonist-specific voltage dependence of M2 muscarinic receptors modulates the deactivation of the acetylcholine-gated K^+ current (I_{KACh}). *Eur J Physiol.* 2016;468:1207–1214.

101. Ruland JG, Kirchhofer SB, Klindert S, Bailey CP, Bünemann M. Voltage modulates the effect of μ-receptor activation in a ligand-dependent manner. *Brit J Pharmacol.* 2020;177(15):3489–3504.

102. Rinne A, Mobarec JC, Mahaut-Smith M, Kolb P, Bünemann M. The mode of agonist binding to a G protein-coupled receptor switches the effect that voltage changes have on signaling. *Sci Signal.* 2015;8(401):ra110. https://doi.org/10.1126/scisignal.aac7419.

CHAPTER SEVEN

Recent advances in calcium-sensing receptor structures and signaling pathways

Caroline M. Gorvin[a,b,]*

[a]Centre of Membrane Proteins and Receptors (COMPARE), Universities of Birmingham and Nottingham, Birmingham, United Kingdom
[b]Institute of Metabolism and Systems Research (IMSR) and Centre for Endocrinology, Diabetes and Metabolism, Birmingham Health Partners, University of Birmingham, Birmingham, United Kingdom
*Corresponding author: e-mail address: c.gorvin@bham.ac.uk

Contents

Abstract

The calcium-sensing receptor (CaSR) is a class C GPCR that has a fundamental role in extracellular calcium homeostasis by regulating parathyroid hormone release and urinary calcium excretion. Germline mutations in the receptor cause disorders of calcium homeostasis and studies of the functional effects of these mutations has facilitated understanding of CaSR signaling and how allosteric modulators affect these responses. In the past year, five cryo-EM structures of the near full-length CaSR have been published, demonstrating how agonist-binding transmits changes in the CaSR extracellular domain to the transmembrane region to activate G proteins, and how allosteric modulators affect these structural dynamics. Additionally, several recent studies have identified CaSR interacting proteins that regulate CaSR signaling and trafficking and contribute to understanding how the receptor achieves rapid and diverse physiological responses.

The calcium–sensing receptor (CaSR) is a class C GPCR, which upon extracellular Ca^{2+} (Ca^{2+}_e) binding activates signaling pathways that regulate parathyroid hormone (PTH) release and urinary calcium excretion.

Progress in Molecular Biology and Translational Science, Volume 195
ISSN 1877-1173
https://doi.org/10.1016/bs.pmbts.2022.06.014
121

Germline mutations in the receptor cause disorders of calcium homeostasis. Inactivating mutations cause familial hypocalciuric hypercalcemia type-1 (FHH1), and rarely cause the potentially fatal neonatal severe hyperparathyroidism (NSHPT); while activating CaSR mutations cause autosomal dominant hypocalcemia type-1 (ADH1).[1] Additionally, inactivating mutations in proteins that CaSR uses to signal, G-protein-a11 (Ga11), and is endocytosed, adaptor protein-2 σ-subunit (AP2σ), cause FHH; whereas activating Ga11 mutations cause ADH2.[1,2] The positive allosteric modulator (PAM), cinacalcet, which was the first allosteric GPCR drug to market, has been used to treat NSHPT and symptomatic hypercalcemia, while negative allosteric modulators (NAMs), designed to treat osteoporosis, could be repurposed in conditions such as asthma, based on favorable results in animal models.[3,4] This review summarizes CaSR studies from the last two years, focusing primarily on cryo-EM structures that have revealed receptor activation mechanisms, and also considers new insights into CaSR signaling that could explain how the receptor achieves rapid and diverse physiological responses.

1. Advances in understanding CaSR activation mechanisms

Two crystal structures of the homodimeric CaSR extracellular domain (ECD) were published in 2016 that demonstrated how agonist binding induces ECD closure to initiate receptor activation, and how co-agonists may facilitate this.[5,6] In 2021, several near full-length cryo-EM structures, incorporating the ECD, 7 transmembrane region (7TM), and extracellular and intracellular loops (ECLs and ICLs), were published by independent groups[7–10] (Table 1). These structures confirmed that two CaSR protomers bind in a side-by-side configuration, with the ECD comprising a bilobed venus flytrap (VFT) domain with a cysteine-rich region that connects the VFT to the 7 transmembrane (TM) region (Fig. 1A). Insights from these structures, alongside FRET analyses, has provided critical information on how agonist-binding transmits changes in the ECD, to the transmembrane region to activate G proteins, and how allosteric modulators affect CaSR structural dynamics.[7–10]

Table 1 Location of ligand-binding sites on the different CaSR structures.

Reference	Type of structure	Ca site 1 (Top of LB1)	Ca site 2 (LB1–LB2 cleft)	Ca site 3 (LB1–LB2 cleft)	Ca site 4 (LB2-CRD homodimer interface)	L-Trp/TNCA (LB1–LB2 cleft)	Other binding sites (anions)
Geng et al.[5]	Human crystal structure ECD	I81, N82, L91, S84, L87	T100, N102, T145, G146	S302, S303	E231, D234, G557	W70, T145, S147, A168, S170, Y218, E297, A298	PO_4^{3-} Site 1 — R66, R69, W70, R415, S417
							PO_4^{3-} Site 2 — E191, H192, T195, E229, K517, R520
	Inactive (Ca^{2+}/PO_4^{3-} bound), active (L-Trp/Ca^{2+} bound)						SO_4^{2-} Site 1 — R62, Y63
							SO_4^{2-} Site 2 and 3 — R66, R69, W70 T412, H413, R415, S417
Zhang et al.[6]	Human crystal structure ECD	I81, S84, L87, L88 (described with Mg^{2+} bound)			E228, E231, S240, E241 (described with Mg^{2+} bound)	N64, R66, G67, W70, N102, T145, S147, A168, S170, I187, S218, S272, E297, A298 S302, I416	HCO_3^- — R66, R69, W70, I416, S417
	Active (TNCA, Mg^{2+} bound)				E228, E229, E232 (described with Gd^{3+} bound)		

Continued

Table 1 Location of ligand-binding sites on the different CaSR structures.—cont'd

Reference	Type of structure	Ca site 1 (Top of LB1)	Ca site 2 (LB1-LB2 cleft)	Ca site 3 (LB1-LB2 cleft)	Ca site 4 (LB2-CRD homodimer interface)	L-Trp/TNCA (LB1-LB2 cleft)	Other binding sites (anions)
Ling et al.[7]	Human cryo-EM ECD-TMD		Shared site with L-Trp		E231, D234, G557	W70, T145, S147, S170, Y218, E297	
	Three states: open-open (inactive), open-closed (L-Trp bound), active closed-closed (L-Trp/Ca^{2+} bound)						
Gao et al.[9]	Human cryo-EM ECD-TMD	I81, S84, L87, L88, L91			E231, D234, G557	R66, S147	
	Open-closed (L-Trp bound inactive), active closed-closed (L-Trp/Ca^{2+} bound)						
	PAM (etelcalcetide, evocalcet, cinacalcet)						
	NAM (NPS-2143)						
Park et al. 2021[10]	Human cryo-EM ECD-TMD	I81, N82, S84, L87, L88, L91	L37, F38, T100, N102, T145, Y167	A298, S302	E231, D234, G557, E558	R66, W70, T145, S147, A168, S170, Y218, A296, E297	PO$_4^{3-}$ R66, R69, W70, R415, S417, I418
	Active (TNCA/Ca^{2+} bound)						
	PAM (R-568)						
	NAM (NPS-2143, inactive, open-open)						

Wen et al.[11]	Chicken cryo-EM ECD-TMD	T100, T145, G146	R66, S302, S303	E231, D234, G556	W70, T145, S147, A168, S170, Y218 E297, A298	Cl⁻	R66, R69, W70, L413, R414, I415
	Inactive (open-open)						
	Active (L–Trp/Ca^{2+} bound)						
	PAM (evocalcet with Ca^{2+} and L–Trp)						
	NAM (NPS-2143 with Ca^{2+} and L–Trp)						
Liu et al.[12]	FRET	Site 1 and 2—S170, D190, Q193, D216, Y218, S272, D275, E297					

Ca^{2+} binding site nomenclature is based on that defined by Geng et al.[5]

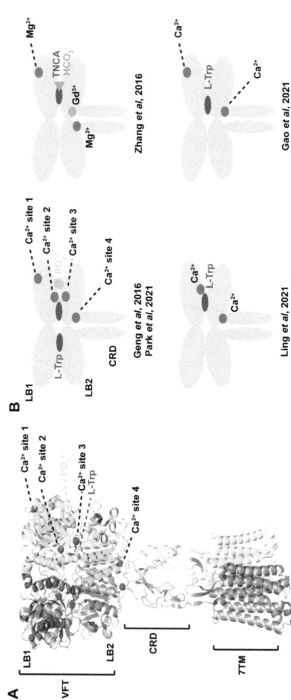

Fig. 1 Location of agonist binding sites in different CaSR structures. (A) Structure of the active full-length CaSR from Park et al. (PDB:7SIM).[10] The extracellular domain (ECD) comprises a venus flytrap (VFT) domain with two lobes (lobe 1, LB1 and lobe 2, LB2), and a cysteine-rich region (CRD) that connects the VFT to the 7 transmembrane (7TM) region. Consistent with a previous crystal structure of the ECD,[5] 4 Ca^{2+} binding sites, and a site for PO_4^{3-} and L-Trp were identified. (B) Cartoons showing the locations of the binding sites for cations and anions in different CaSR crystal and cryo-EM structures.[5–7,9,10]

2. Identification of agonist binding sites

The two previously published CaSR ECD structures identified a number of binding sites for Ca^{2+}, L–Trp and anions. Geng et al. identified 4 Ca^{2+} binding sites: site 1 at the top of Lobe 1 (LB1), site 2 and 3 at the LB1-LB2 VFT cleft, with L–Trp close to site 3, and site 4 at the LB2-CRD homodimeric interface (Fig. 1B, Table 1).[5,6] Zhang et al. purified their ECD structure with Mg^{2+} and identified three metal binding sites and an L–Trp site that shared similarities with these four Ca^{2+} binding sites (Fig. 1C, Table 1).[5,6] A cell-free FRET study later questioned the feasibility of these Ca^{2+} binding sites, as mutation of multiple residues within these regions had no effect on CaSR activity, while other sites were found to occupy anions.[12] Instead, two different Ca^{2+} binding sites were identified at the VFT cleft, close to the L–Trp binding site, that occupy water molecules in the ECD structures, and had previously been predicted in computational studies to bind Ca^{2+} ions[5,6,12,13] (Table 1).

The first cryo-EM structure identified only two Ca^{2+} binding sites in each CaSR protomer: one within the LB1-LB2 cleft adjacent to the L–Trp binding site and one at the LB2-CRD interface,[7] similar to site 2 and 3 of the ECD structures[5] (Fig. 1B, Table 1). The second cryo-EM structure also identified two Ca^{2+} binding sites, one at the top of LB1 and the second in the LB2-CRD interface, corresponding to site 1 and 4[5,6,9] (Fig. 1B, Table 1). All four of the original Ca^{2+} binding sites identified in ECD structures were detected in a later study that described a symmetric conformation[10]; while the chicken CaSR (ggCaSR) cryo-EM structure, which has 80% identity with human CaSR, identified three Ca^{2+} binding sites, sites 2–4 (Fig. 1B, Table 1). Therefore, while these structures do not identify precisely the same sites, there are commonalities, which combined with mutagenesis studies and knowledge of disease–causing mutations, indicates that residues surrounding all four sites likely play a role in agonist binding or receptor activation.

Amino acids such as L–Trp can also increase CaSR activity, and a binding site for L–Trp (or its derivative TNCA) has been identified within the VFT cleft, close to a Ca^{2+} binding site.[5,6] An L–Trp binding site was also observed in the human and chicken cryo-EM structures in a similar site.[8–11] Several residues within this L–Trp binding region are sites of inactivating CaSR mutations,[1,7,11] demonstrating the importance of amino acid binding for CaSR activity. In one study it was suggested that Ca^{2+} and L–Trp may synergistically activate CaSR,[11] although this active state was established using a nanobody, which could sterically hinder movement, and may not reflect the nanobody-free active state. Other structural and FRET studies consistently show that L-Trp modulates CaSR activation, but cannot activate the receptor in the absence of Ca^{2+}, and indicate that mutations in the L-Trp binding site modify L-Trp allosteric effects rather than activate the receptor in the absence of Ca^{2+}.[9,12]

Several structures also identified anion binding sites[10–12] (Table 1). These sites may bind phosphate ions,[10,11] which have previously been shown to inhibit the receptor,[14] or chloride ions, which increase the potency of Ca^{2+} in TR-FRET assays[12] and have been shown to activate other class C GPCRs.[15] Indeed, mutation of a conserved residue that binds Cl^-, Thr100, impairs agonist-induced activation of the CaSR and mGluR4, and is associated with hyperparathyroidism in patients with the Thr100Ile mutation.[1,12,15]

3. Activation of the CaSR

The first cryo-EM study captured the CaSR VFT in three conformations: open-open, open-closed and closed-closed.[7,9,10] However, there are disagreements regarding the transition from inactive to active CaSR structures, with supporters of the three state model suggesting: the open-open state represents fully inactive, while the intermediate open-closed VFT state is adopted upon L-Trp binding, to more readily facilitate conformational changes required for switching to the fully active closed-closed state.[7,9] In contrast, Park et al. described two VFT states: an open-open inactive and a closed-closed active state, with the latter binding Ca^{2+}, L-Trp, and anions (likely phosphate or chloride),[10] that were similarly identified in the ggCaSR structure.[11] The previously observed open-closed state was hypothesized to occur when ambient amino acids are present during experimental preparations.[10]

There are also disparities between predicted TMD structural conformations. One study suggested that CaSR adopts an asymmetric conformation stabilized by interactions across the CRD-ECL2-ECL3.[9] Multiple disease-causing mutations have been described in these regions, highlighting their importance in CaSR structure-function.[1,9] Consistent with an asymmetric conformation, ECL2-ECL3 interactions and PAM poses are distinct between the two protomers.[9] This asymmetry was proposed to favor the engagement of G proteins sequentially, as shown previously for other family C GPCR dimers.[16] However, shortly thereafter, another cryo–EM structure described activation by homodimeric human CaSR with a symmetric conformation within the ECD and TMD,[10] similar to that already described for *gg*CaSR.[11] A structure of the CaSR in complex with a G protein will likely provide more information on whether the receptor adopts a symmetric or asymmetric conformation.

Despite the noted differences, all recent CaSR structure papers agree that the CaSR has multiple interfaces between the VFT (LB1 and LB2), CRD and TMD of each subunit, and propose a relay of rotations within inter-subunit interfaces are required to form the fully activated receptor.[7,9,10] The first rotation involves the long arm-like loop in the LB1 regions, which stretches from one subunit to the other at the top of the receptors (between Leu51 in the loop and Phe444 and Trp458 in the helix), and has previously been described in the CaSR–ECD structures.[5,6] Interactions and conformational changes at this interface are thought to initiate domain twisting in the homodimer, which brings the LB2 regions and CRDs into closer proximity to expand the homodimeric interactions along LB2-LB2 and CRD-CRD.[7,8,11] Intra-subunit hydrophobic interactions between the CRD and ECL2 are important for propagating the conformational changes from the ECD to TMD.[7,8] A number of mutations have been identified in ECL2, including two activating mutations in Phe832, which plays a critical role in the ECL2-CRD interaction[1,7] and has been shown to bind to the NAM NPS-2143.[17]

The homodimeric interface is then expanded to the TMD. In the open-closed or inactive conformation, TM5-TM6 face each other, but do not contact[9,10] (Fig. 2A). However, on activation, TM6 helices move closer together to form a TM6-TM6 interface, similar to other class C GPCRs,[18,19] and consistent with previous findings from cysteine cross-linking and a CaSR FRET biosensor[12] (Fig. 2B). ADH1-associated activating CaSR mutations are known to cluster around TM6.[1,9–11] Movement of the TM6 helices is facilitated by the formation of a kink at Pro823 of TM6 upon activation[9,10] (Fig. 2B). Residues within this breakpoint region, such as Phe821, undergo dramatic changes upon receptor activation, resulting in reorientation from pointing towards the core of the TM bundle to facing the dimer interfacial region. Mutations in Ala824 and Phe821, which facilitate TM6-kink formation and TM6 rotation, cause ADH1.[1,9] Additionally, re-orientating of the TMDs from TM5-TM6 to TM6-TM6 may be propagated by semi-rigidity in the CRD, which allows the small rotation of the LB2 domains to propagate large-scale transitions to the TMDs.[11]

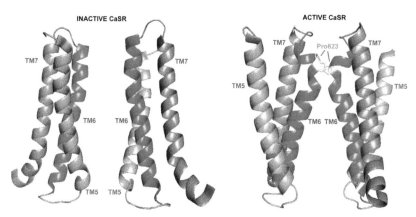

Fig. 2 Conformational changes in the transmembrane region required for activation. (A) Location of TM5 (green), TM6 (blue), TM7 (red) in the inactive CaSR structure (PDB:7SIN). TM5 and TM6 face each other, but do not contact. (B) Location of TM5-7 in the active CaSR structure (PDB:7SIM). On activation, the TM6 helices move closer together to form a TM6–TM6 interface. Movement of the TM6 helices is facilitated by the formation of a kink at Pro823 (shown in yellow).

The divergence of TM6-TM6 allows cholesterol containing substances to bind to the intracellular and extracellular TM complex, which may be important for stabilizing the TM6-TM6 interface, as well as CaSR clustering within cholesterol-rich caveolae previously described in parathyroid cells.[9,10,20] Finally, structural studies have confirmed the presence of a TM6 toggle switch in CaSR, which has previously been identified in mGluRs,[21] and drives the opening of the TM6 cytoplasmic ends to facilitate G protein binding.[9]

4. PAM/NAM binding

Cryo-EM studies have also revealed details of how allosteric modulators bind to CaSR and affect receptor activation. Etelcalcetide binds within the ECD, while cinacalcet and evocalcet bind to the TMD.[12] In the PAM-bound conformation, homodimer interactions between CaSR TMDs are more extensive, thus stabilizing the active conformation.[10] Two copies of the PAM etelcalcetide bind to the CaSR homodimer within the CRD interface with interactions formed between CaSR-Cys482 and the N-terminus of etelcalcetide, and Glu228 and Glu241 on the opposite protomer binding to the C-terminus of etelcalcetide.[9] This stabilizes the active closed-closed state of the ECD to facilitate activation. The R-568 binding site is similarly located near the extracellular TMD and is framed by residues from TM2, 3, 5, 6 and 7, with one PAM molecule bound to each protomer.[10] Its binding is compatible with the TM6 helix conformation found in the active state CaSR.[10] In the ggCaSR structure, evocalcet binding results in a closer distance between the extracellular ends of the two TM6s and creates a greater distance between the cytoplasmic ends.[11]

Three studies examined binding of the NAM NPS-2143 to CaSR. Gao et al. showed that the 7TMs are symmetric in the NAM-bound state, with distinct interfaces when compared to the Ca^{2+}-L-Trp bound state.[9] Park et al., described NPS-2143 binding to a site similar to that identified for PAMs.[10] In this structure the NAM stabilizes an intact TM6 conformation in which interactions are formed with the Phe821 residue in its inward facing conformation, and strengthens a TM6 helicity that sterically hinders activation.[10] In the ggCaSR a cooperative binding mode was described between the two NPS-2143 molecules on each protomer, similar to that described in previous studies of human CaSR.[22,23] Structures of combined G protein and PAM/NAM-bound CaSR are likely to provide more insights into whether allosteric modulators support asymmetric or symmetric conformations.

5. New insights into CaSR signaling

Several publications published in the last 2 years have advanced understanding of CaSR signaling. CaSR responds to minute changes in Ca^{2+}_e by rapidly mobilizing a near-membrane pool of receptors to the plasma membrane in a phenomenon named agonist-driven insertional signaling (ADIS).[24] A recent proteomics study that investigated agonist-stimulated CaSR interactions has provided insight into how the receptor may be able to respond rapidly to changes in extracellular calcium.[25] These studies confirmed that rapid release of Ca^{2+} from the ER occurs upon CaSR activation, and pathway analysis identified an enrichment of receptor interactions with proteins of the ER membrane.[25] Specifically, Ca^{2+}_e induced an interaction between CaSR and the vesicle-associated membrane protein-associated A (VAPA), which regulates anterograde trafficking from the ER to the Golgi, and glucose-regulated protein-78 (GRP78), a chaperone involved in ER trafficking. These findings suggest that VAPA and GRP78 may have a role in rapid CaSR responses to Ca^{2+}_e by controlling ER protein processing and surface expression.[25]

Interactions between CaSR and Homer1, a scaffold protein, have also been shown to regulate CaSR expression. CaSR, mTORC2 and Homer1 interact to stimulate AKT phosphorylation and promote differentiation and survival of osteoblasts.[26] Studies have now shown that Homer1 may act as a chaperone for CaSR.[27] In human and mouse osteoblasts protein levels of CaSR and Homer1 were shown to be interdependent, and colocalize at peri-nuclear sites.[27] Homer1 has previously been shown to regulate type-1 metabotropic glutamate receptors,[28] and thus may have a more general role in regulating expression and signaling of class C GPCRs.

Previous studies had indicated that while CaSR predominantly signals via $G\alpha_{q/11}$ and $G\alpha_{i/o}$, activation of other G protein families can occur in some cell types.[29,30] Recent studies using a NanoBiT G protein dissociation assay to test 12 $G\alpha$ subunits confirmed that CaSR activates all 4 G protein families, and that tissue-specific signaling is governed by expression of individual $G\alpha$ proteins, rather than G protein bias.[31] Thus, in parathyroid tissue, in which CaSR has its dominant role, $G\alpha_{11}$ and $G\alpha_{i1}$ are most abundant, demonstrating why activation of CaSR increases intracellular calcium (Ca^{2+}_i) and reduces cAMP in parathyroid cells. CaSR activation of $G\alpha_{12/13}$ was slow, providing a reason why CaSR-mediated $G\alpha_{12/13}$ signaling is rarely reported and appears to be tissue-specific.[32] Consistent with preliminary studies,[2] $G\alpha_{11}$ was expressed at significantly greater levels than $G\alpha_q$ in human parathyroid tissue, explaining why $G\alpha_q$ cannot entirely compensate for inactivating $G\alpha_{11}$ mutations and patients develop hypercalcemia.[31]

Finally, two recent studies have shown how the CaSR may perform diverse functions in tissues that regulate calcium homeostasis. The enzyme 1α-hydroxylase, encoded by the CYP27B1 (cytochrome P450 family 27, subfamily B member 1) gene, catalyzes the synthesis of 1,25-dihydroxyvitamin D3, and is modulated both positively and negatively by Ca^{2+}_e depending on the tissue.[33] Studies in HEK293-CaSR cells have shown that CaSR can control CYP27B1 expression in a biphasic manner, with 0.5 and 3 mM Ca^{2+}_e inducing CYP27B1 luciferase activity, and $Ca^{2+}_e > 5$ mM suppressing CYP27B1. The suppression of CYP27B1 at high Ca^{2+}_e is mediated by phosphorylation of Thr888 and activation of two signaling pathways (PKC and ERK).[33] The authors hypothesized that the inhibitory effect of $Ca^{2+}_e > 5$ mM may be important in the renal proximal tubule, which is under tonic stimulation from PTH, whereas activation at low Ca^{2+}_e may be important in parathyroid and skeletal tissues, where Ca^{2+} positively regulates 1α-hydroxylase.[33]

Another kidney-expressed gene that is regulated by CaSR is the tight junction protein Claudin-14, and a recent study has shown mechanisms by which CaSR may regulate urinary calcium excretion. Activation of CaSR by cinacalcet increased Claudin-14 luciferase activity, which was suppressed when PLC, p38 and Sp1 were inhibited, indicating these three proteins are involved in the signaling pathway.[34] Sp1 is a transcriptional repressor of Claudin-14, and CaSR was shown to reduce its phosphorylation, thus increasing Claudin-14 activity in the thick ascending limb.[34] Finally, over-expression of an FHH1 mutation reduced Claudin-14 reporter activity, while ADH1 mutants increased reporter activity, which could affect paracellular calcium transport at the thick ascending limb, and contribute to inappropriate urinary calcium excretion in these patients.[34] Thus, CaSR activation increases Claudin-14 expression via a PLC-stimulated p38 pathway.

6. Summary

The elucidation of several near full-length CaSR structures have advanced understanding of how agonists and allosteric modulators bind and induce conformational changes necessary for receptor activation. Additional structures of CaSR in the G protein bound state are likely to arise in the near future and could resolve the disparities between the current cryo-EM structures. New insights into the mechanisms by which interacting proteins affect CaSR activity has improved understanding of how CaSR may perform rapid and tissue-specific responses. Further studies in different cell types, using dynamic rather than static assays, is likely to enhance understanding of CaSR signaling and trafficking mechanisms. Together, new structural and signaling insights could be exploited for the rational design of CaSR targeting drugs with fewer side effects.

Declaration of interest

The author declares that there is no conflict of interest that could be perceived as prejudicing the impartiality of this review.

Funding

CMG is in receipt of research funding in the form of an Academy of Medical Sciences Springboard Award (Ref: SBF004|1034), which is supported by the British Heart Foundation, Diabetes UK, the Global Challenges Research Fund, the Government Department of Business, Energy and Industrial Strategy and the Wellcome Trust.

References

1. Gorvin CM. Molecular and clinical insights from studies of calcium-sensing receptor mutations. *J Mol Endocrinol*. 2019;63(2):R1–R16. https://doi.org/10.1530/JME-19-0104.

2. Nesbit MA, Hannan FM, Howles SA, et al. Mutations affecting G-protein subunit alpha11 in hypercalcemia and hypocalcemia. *N Engl J Med*. 2013;368(26):2476–2486. https://doi.org/10.1056/NEJMoa1300253.

3. Yarova PL, Huang P, Schepelmann MW, et al. Characterization of negative allosteric modulators of the calcium-sensing receptor for repurposing as a treatment of asthma. *J Pharmacol Exp Ther*. 2021;376(1):51–63. https://doi.org/10.1124/jpet.120.000281.

4. Diao J, DeBono A, Josephs TM, et al. Therapeutic opportunities of targeting allosteric binding sites on the calcium-sensing receptor. *ACS Pharmacol Transl Sci*. 2021;4 (2):666–679. https://doi.org/10.1021/acsptsci.1c00046.

5. Geng Y, Mosyak L, Kurinov I, et al. Structural mechanism of ligand activation in human calcium-sensing receptor. *Elife*. 2016;5. https://doi.org/10.7554/eLife.13662.

6. Zhang C, Zhang T, Zou J, et al. Structural basis for regulation of human calcium-sensing receptor by magnesium ions and an unexpected tryptophan derivative co-agonist. *Sci Adv*. 2016;2(5), e1600241. https://doi.org/10.1126/sciadv.1600241.

7. Ling S, Shi P, Liu S, et al. Structural mechanism of cooperative activation of the human calcium-sensing receptor by Ca(2+) ions and L-tryptophan. *Cell Res*. 2021;31 (4):383–394. https://doi.org/10.1038/s41422-021-00474-0.

8. Chen X, Wang L, Cui Q, et al. Structural insights into the activation of human calcium-sensing receptor. *Elife*. 2021;10. https://doi.org/10.7554/eLife.68578.

9. Gao Y, Robertson MJ, Rahman SN, et al. Asymmetric activation of the calcium-sensing receptor homodimer. *Nature*. 2021;595(7867):455–459. https://doi.org/10.1038/s41586-021-03691-0.

10. Park J, Zuo H, Frangaj A, et al. Symmetric activation and modulation of the human calcium-sensing receptor. *Proc Natl Acad Sci USA*. 2021;118(51). https://doi.org/10.1073/pnas.2115849118.

11. Wen T, Wang Z, Chen X, et al. Structural basis for activation and allosteric modulation of full-length calcium-sensing receptor. *Sci Adv*. 2021;7(23). https://doi.org/10.1126/sciadv.abg1483.

12. Liu H, Yi P, Zhao W, et al. Illuminating the allosteric modulation of the calcium-sensing receptor. *Proc Natl Acad Sci USA*. 2020;117(35):21711–21722. https://doi.org/10.1073/pnas.1922231117.

13. Silve C, Petrel C, Leroy C, et al. Delineating a Ca2+ binding pocket within the venus flytrap module of the human calcium-sensing receptor. *J Biol Chem*. 2005;280 (45):37917–37923. https://doi.org/10.1074/jbc.M506263200.

14. Centeno PP, Herberger A, Mun HC, et al. Phosphate acts directly on the calcium-sensing receptor to stimulate parathyroid hormone secretion. *Nat Commun*. 2019; 10(1):4693. https://doi.org/10.1038/s41467-019-12399-9.

15. Tora AS, Rovira X, Dione I, et al. Allosteric modulation of metabotropic glutamate receptors by chloride ions. *FASEB J*. 2015;29(10):4174–4188. https://doi.org/10.1096/fj.14-269746.

16. Hlavackova V, Goudet C, Kniazeff J, et al. Evidence for a single heptahelical domain being turned on upon activation of a dimeric GPCR. *EMBO J*. 2005;24(3):499–509. https://doi.org/10.1038/sj.emboj.7600557.

17. Leach K, Gregory KJ, Kufareva I, et al. Towards a structural understanding of allosteric drugs at the human calcium-sensing receptor. *Cell Res*. 2016;26(5):574–592. https://doi.org/10.1038/cr.2016.36.

18. Koehl A, Hu H, Feng D, et al. Structural insights into the activation of metabotropic glutamate receptors. *Nature*. 2019;566(7742):79–84. https://doi.org/10.1038/s41586-019-0881-4.

19. Shaye H, Ishchenko A, Lam JH, et al. Structural basis of the activation of a metabotropic GABA receptor. *Nature.* 2020;584(7820):298–303. https://doi.org/10.1038/s41586-020-2408-4.

20. Kifor O, Diaz R, Butters R, Kifor I, Brown EM. The calcium-sensing receptor is localized in caveolin-rich plasma membrane domains of bovine parathyroid cells. *J Biol Chem.* 1998;273(34):21708–21713. https://doi.org/10.1074/jbc.273.34.21708.

21. Seven AB, Barros-Alvarez X, de Lapeyriere M, et al. G-protein activation by a metabotropic glutamate receptor. *Nature.* 2021;595(7867):450–454. https://doi.org/10.1038/s41586-021-03680-3.

22. Gregory KJ, Kufareva I, Keller AN, et al. Dual action calcium-sensing receptor modulator unmasks novel mode-switching mechanism. *ACS Pharmacol Transl Sci.* 2018;1(2):96–109. https://doi.org/10.1021/acsptsci.8b00021.

23. Jacobsen SE, Gether U, Brauner-Osborne H. Investigating the molecular mechanism of positive and negative allosteric modulators in the calcium-sensing receptor dimer. *Sci Rep.* 2017;7:46355. https://doi.org/10.1038/srep46355.

24. Grant MP, Stepanchick A, Cavanaugh A, Breitwieser GE. Agonist-driven maturation and plasma membrane insertion of calcium-sensing receptors dynamically control signal amplitude. *Sci Signal.* 2011;4(200). https://doi.org/10.1126/scisignal.2002208, ra78.

25. Gorkhali R, Tian L, Dong B, et al. Extracellular calcium alters calcium-sensing receptor network integrating intracellular calcium-signaling and related key pathway. *Sci Rep.* 2021;11(1):20576. https://doi.org/10.1038/s41598-021-00067-2.

26. Rybchyn MS, Islam KS, Brennan-Speranza TC, et al. Homer1 mediates CaSR-dependent activation of mTOR complex 2 and initiates a novel pathway for AKT-dependent beta-catenin stabilization in osteoblasts. *J Biol Chem.* 2019;294(44):16337–16350. https://doi.org/10.1074/jbc.RA118.006587.

27. Rybchyn MS, Brennan-Speranza TC, Mor D, et al. The mTORC2 regulator homer1 modulates protein levels and sub-cellular localization of the CaSR in osteoblast-lineage cells. *Int J Mol Sci.* 2021;22(12). https://doi.org/10.3390/ijms22126509.

28. Mao L, Yang L, Tang Q, Samdani S, Zhang G, Wang JQ. The scaffold protein Homer1b/c links metabotropic glutamate receptor 5 to extracellular signal-regulated protein kinase cascades in neurons. *J Neurosci.* 2005;25(10):2741–2752. https://doi.org/10.1523/JNEUROSCI.4360-04.2005.

29. Mamillapalli R, VanHouten J, Zawalich W, Wysolmerski J. Switching of G-protein usage by the calcium-sensing receptor reverses its effect on parathyroid hormone-related protein secretion in normal versus malignant breast cells. *J Biol Chem.* 2008;283(36):24435–24447. https://doi.org/10.1074/jbc.M801738200.

30. Huang C, Hujer KM, Wu Z, Miller RT. The Ca2+-sensing receptor couples to Galpha12/13 to activate phospholipase D in Madin-Darby canine kidney cells. *Am J Physiol Cell Physiol.* 2004;286(1):C22–C30. https://doi.org/10.1152/ajpcell.00229.2003.

31. Abid HA, Inoue A, Gorvin CM. Heterogeneity of G protein activation by the calcium-sensing receptor. *J Mol Endocrinol.* 2021;67(2):41–53. https://doi.org/10.1530/JME-21-0058.

32. Pi M, Spurney RF, Tu Q, Hinson T, Quarles LD. Calcium-sensing receptor activation of rho involves filamin and rho-guanine nucleotide exchange factor. *Endocrinology.* 2002;143(10):3830–3838. https://doi.org/10.1210/en.2002-220240.

33. Huang A, Binmahfouz L, Hancock DP, Anderson PH, Ward DT, Conigrave AD. Calcium-sensing receptors control CYP27B1-luciferase expression: transcriptional and posttranscriptional mechanisms. *J Endocr Soc.* 2021;5(9). https://doi.org/10.1210/jendso/bvab057, bvab057.

34. Lee JJ, Alzamil J, Rehman S, Pan W, Dimke H, Alexander RT. Activation of the calcium sensing receptor increases claudin-14 expression via a PLC -p38-Sp1 pathway. *FASEB J.* 2021;35(11), e21982. https://doi.org/10.1096/fj.202002137RRR.

CHAPTER EIGHT

Structural insights into promiscuous GPCR-G protein coupling

Ángela Carrión-Antolí, Jorge Mallor-Franco, Sandra Arroyo-Urea, and Javier García-Nafría*

Institute for Biocomputation and Physics of Complex Systems (BIFI) and Laboratorio de Microscopías Avanzadas (LMA), University of Zaragoza, Zaragoza, Spain
*Corresponding author: e-mail address: jgarcianafria@unizar.es

Contents

Abstract

G protein-coupled receptors (GPCRs) transduce extracellular signals across biological membranes by activating heterotrimeric G$\alpha\beta\gamma$ proteins. There are 16 different human Gα proteins grouped into four families (G$_S$, G$_{I/O}$, G$_{q/11}$ and G$_{12/13}$), each one activating different signaling cascades. Around 50% of non-olfactory GPCRs activate more than one type of Gα proteins with different efficacy and kinetics, triggering a fingerprint-like signaling profile. In this chapter we review the GPCR-G protein promiscuity landscape and discuss recent structures of GPCRs coupled to different Gα proteins. Overall, the size and shape of the intracellular cavity (determined by the extent of outward movement of TM6) is maintained when the receptor is coupled to different Gα proteins, and is determined by the type of primary Gα coupling. The "sub-optimal" secondary Gα coupling is further supported by interactions with the intracellular loops, with ICL2 and ICL3 having a relevant role in promiscuous couplings

Progress in Molecular Biology and Translational Science, Volume 195
ISSN 1877-1173
https://doi.org/10.1016/bs.pmbts.2022.06.015
137

1. Introduction

G protein-coupled receptors (Gpcrs) form the largest family of membrane receptors (>800 members in humans) and recognize a staggering amount of extracellular signals (~1000) that range from subatomic particles (photons) to macromolecules.[1] Their high versatility in signal detection as well as their ubiquitous distribution involves GPCRs in a wide variety of (patho-)physiological processes as well as being highly prolific therapeutic targets.[2,3] Upon detection of stimuli GPCRs transduce the information across biological membranes into the intracellular milieu where they recruit and activate heterotrimeric $G\alpha\beta\gamma$ proteins and arrestins.[4] Heterotrimeric $G\alpha\beta\gamma$ proteins are the primary route for signal transduction which, upon coupling and activation by GPCRs, dissociate into the $G\alpha$ and $G\beta\gamma$ subunits triggering an array of signaling cascades through various effectors (e.g., adenylate cyclase, phospholipase C...) and secondary messengers (cAMP, Ca^{2+}, DAG...) that lead to a cell-specific response.[5,6] The nature of the activated signaling cascade depends mainly on the type of $G\alpha$ protein. In humans, there are 16 genes that code for distinct $G\alpha$ proteins organized into four families: G_S (G_{olf} and G_S), $G_{i/O}$ (G_{i1}, G_{i2}, G_{i3}, G_O, G_z, G_{t1}, G_{t2} and G_{gust}), $G_{q/11}$ (G_q, G_{11}, G_{14}, G_{15}) and $G_{12/13}$(G_{12} and G_{13}). The main signaling routes initiated from the different $G\alpha$ proteins are well established: $G\alpha_S$ activates adenylate cyclase and promote the formation of cAMP, $G\alpha_{i/O}$ inhibits the formation of cAMP, $G\alpha_{q/11}$ activates phospholipase C and consequently calcium signaling, and $G\alpha_{12/13}$ activates Rho A GTPases. Although differential expression and sub-cellular compartmentalization can influence the ability of certain GPCRs to activate specific Gproteins,[7–10] it is known that many GPCRs and G proteins can be highly expressed in a single cell-type.[11,12] Hence, a major contributor to GPCR-G protein selectivity is likely to be the set of specific interactions between GPCRs and G proteins. GPCRs can be specific, coupling to and activating a single type of $G\alpha\beta\gamma$ heterotrimer, or have different degrees of promiscuity, where additional primary and/or secondary couplings to other $G\alpha$ proteins occur. Promiscuous couplings increase the complexity of GPCR signaling, activating different $G\alpha$ proteins with different strengths (efficacies) and kinetics yielding a fingerprint-like profile.[13,14] Such a complex signaling is bound to be tunable by receptor environment or distinct endogenous/exogenous agonists through biased agonism/

functional selectivity.[15] In this chapter we review the advances in characterizing promiscuity within the GPCR family and analyze recent structures of GPCRs coupled to different Gα proteins.

2. On the search for the GPCR–G protein *couplome*

A map of the GPCR *couplome* that includes detailed information of which Gα proteins are activated by which GPCRs (with associated efficacy/kinetic information when activated by different agonists) would be of great value, and efforts are directed toward that goal. Information about individual GPCR-G protein couplings is recorded, in a qualitative manner (as primary/secondary couplings), in the IUPHAR/BPS Guide to Pharmacology.[16] This information originates from the literature and is expert-curated. Additionally, the development of robust cellular Bioluminescence Resonance Energy Transfer (BRET) assays that monitor Gαβγ activation has allowed more systematic comparisons of GPCR-G protein couplings.[17–20] Two recent large-scale studies and their quantitative merging and normalization have enhanced our knowledge on the GPCR *couplome*.[21–23] First, Inoue *et al.* used a TGF-α shedding assay in HEK293 cells to study the coupling of the 16 human Gα proteins to 148 non-olfactory receptors. In this case the wild-type G_q was used together with chimeric Gα proteins where the six C-terminal residues were replaced by their corresponding counterparts in other Gα proteins. Additionally, Avet *et al.* used a G protein Effector Membrane Translocator assay (GEMTA) where BRET sensors were used to monitor Gαβγ activation by measuring the translocation of downstream effectors to the plasma membrane. Such an approximation enabled the use of wild-type G proteins and receptors and it was used to determine the GPCR-G protein couplings of 100 GPCRs to the most ubiquitous Gα proteins (G_S, G_{i1}, G_{i2}, G_z, G_{OA}, G_{OB}, G_q, G_{11}, G_{14}, G_{15}, G_{12} and G_{13} and excluding the specific G_{olf}, G_{i3}, G_{t1}, G_{t2} and G_{gust}). A final study merged and normalized the GPCR-G protein couplings from both datasets and joined it to the information in the IUPHAR/BPS Guide to Pharmacology.[22] The final consensus map of GPCR-G protein couplings (deposited in the GPCRdb[24,25]) includes coupling information of 265 non-odorant receptors (67% coverage of non-olfactory GPCRs). Several insights about GPCR-G protein promiscuity can be learnt from this data. First, Gα protein promiscuity is a common feature in GPCRs, with ~50% of the receptors (130/256) coupling to two or more types of Gα proteins (G_S, $G_{i/O}$, $G_{q/11}$ and $G_{12/13}$). Such magnitude of promiscuity is in agreement with previous estimations using the IUPHAR/BPS Guide to Pharmacology.[12] Within the promiscuous receptors, ~64% (83 receptors) have double couplings, ~26% (34 receptors) have triple couplings and 10% (13 receptors) could activate all families of Gα proteins (G_S, $G_{i/O}$, $G_{q/11}$ and $G_{12/13}$). All of the later highly promiscuous receptors are Class A GPCRs. Second, there is generally little coupling selectivity between Gα protein sub-types, i.e., GPCRs that couple to the $G_{i/O}$ family can normally couple to all sub-types of $G_{i/O}$ proteins. This is somewhat expected due to the high sequence similarity between Gα protein sub-types but some GPCRs showed selectivity for a particular sub-type.[22] Since different Gα proteins sub-types have differences in effector engagement selectivity and kinetic profiles,[13] receptor sub-type selectivity can yield relevant differences in functional outcomes.[26,27] Third, promiscuous receptors showed a negative correlation for co-coupling of G_S and $G_{i/O}$ (this is expected since G_S and $G_{i/O}$ have opposite functional effects) but showed a positive correlation for $G_{i/O}$ and $G_{q/11}$ co-coupling, i.e. promiscuous receptors that co-couple to $G_{i/O}$ and $G_{q/11}$ are much more frequent that receptors that co-couple to other Gα protein pairs. Finally, receptors that couple primarily to $G_{i/O}$ are more selective than receptors coupling to G_S and $G_{q/11}$ (in line with previous reports from the literature[18]), while receptors that couple to $G_{12/13}$ tend to couple to other Gα proteins frequently (i.e., selective $G_{12/13}$ coupling is uncommon). Overall, GPCR–G protein promiscuity is ubiquitous and thus, is an important element within GPCR signaling.

3. Structural studies of GPCRs coupled to G proteins

The selectivity mechanisms by which GPCRs couple to specific Gα proteins is a subject of intense research with high-resolution structural determination being a highly valuable tool. Although an initial X-ray crystal structure of a GPCR-G_S complex was determined in 2011,[28] the high requirements of X-ray crystallography has made crystallization of GPCR-G protein complexes an arduous task and alternative approximations have been used.[29,30] The cryo-electron microscopy (cryo-EM) "resolution revolution"[31] made structure determination of GPCR-G protein complexes more accessible.[32] Initial cryo-EM structures of Class B and A GPCRs coupled to a G_S heterotrimer[33] were rapidly followed by structures of GPCRs coupled to $G_{i/O}$[34–37] and $G_{q/11}$ heterotrimers.[38] Since then, cryo-EM structures of GPCRs coupled to different G proteins, arrestins and kinases have been growing exponentially[39] and, as of April 2022, over 200 structures of GPCR-G protein complexes have been deposited in the Protein Data Bank.[40] In general, agonists binding at the extracellular orthosteric site triggers a conformational change in the conserved CWxP, PIF, NPxxY and E/DRY motifs in the receptor that converge at the intracellular cavity where there are rearrangements of TM3, TM6 and TM7 that allow to accommodate the C-terminal α5 of the Gα protein.[41,42] The selectivity barcode between GPCRs and G proteins is still not understood, although it is believed that a three-dimensional epitope presented by the Gα protein is read by the receptor determining successful coupling and activation.[12] The α5 of the Gα protein seems to be the major determinant of specificity since replacement of its outmost C-terminal residues are enough to modify its specificity.[43] However, elements outside the α5 have been shown to have differential contributions in a GPCR-G protein specific manner.[18,44] From the initial cryo-EM structures, distinct modes of engagement that are Gα protein dependent arose.[45,46] First, a trend in the magnitude of TM6 outward swing differentiates between G_S and $G_{i/O}$-$G_{q/11}$ coupling receptors which is wide for G_S coupling receptors (accommodating the bulkier G_Sα5) and narrower for $G_{i/O}$ and $G_{q/11}$ coupling receptors (Fig. 1A). Such movement contributes majorly to the size and shape of the intracellular cavity for the α5 of the Gα protein. As usual with GPCRs, this is only a trend and exceptions to the rule have been reported.[47,48] Second, the angle of insertion of the G protein α5 with respect to the receptor TM3 is larger for $G_{i/O}$ coupled receptors than for

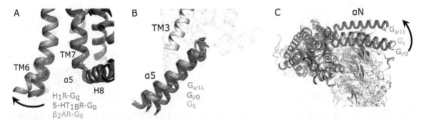

Fig. 1 Engagement modes of different Gα protein. Structures of model GPCRs coupled to G_S (β$_2$ adrenergic receptor, β$_2$AR, PDB 3SN6), $G_{i/O}$ (serotonin 5-HT$_{1B}$ receptor, 5-HT$_{1B}$R, PDB 6G79) and $G_{q/11}$ (Histamine 1 receptor, H$_1$R, PDB 7DFL). Structures are aligned on the receptor and depicted as green (β$_2$AR), blue (H$_1$R) and red (5-HT$_{1B}$R) cartoons. (A) TM6 outward swing in the G_S, G_O and G_{11} coupled receptors. (B) Insertion angle of G_S, $G_{i/O}$ and $G_{q/11}$ α5 into the receptor. (C) Rotation angle of the G_S, $G_{i/O}$ and $G_{q/11}$ with respect to the receptor when view from the extracellular side.

G_S coupling receptors (i.e. $G_{i/O}$ inserts to the receptor more perpendicularly to the membrane than G_S) (Fig. 1B), while an anti-clockwise rotation of the G proteins (as viewed from the extracellular side) tends to be more pronounced for $G_{q/11}$ coupling[38] than for $G_{i/O}$ and G_S coupling (Fig. 1C). Lastly, the insertion and rotation angle is somewhat correlated with the amount of interactions between the intracellular loop 2 (ICL2) and the β1-αN of the Gα protein, with G_q and G_S displaying extensive interactions and $G_{i/O}$ having weaker or absent interactions.[34,45,49]

4. Structures of GPCRs bound to multiple G proteins

There are currently seven receptors whose cryo-EM structures have been determined in the presence of more than one type of Gα proteins. Here we compare, for each receptor, the structures bound to the same agonist but coupled to different Gα proteins. These include: one receptor coupled to G_S and $G_{i/O}$ (GCGR[50]), one receptor coupled to G_S and $G_{q/11}$ (NK$_1$R[51]), one receptor coupled to G_q, G_{i1} and G_S (CCK$_A$R[52,53]) and four receptors coupled to $G_{i/O}$ and $G_{q/11}$ (GSHR,[49,52,54] CCK$_B$R,[55] GPR139[56] and MRGPRX2[57]) (Table 1). Overall, there are six Class A receptors and one Class B receptor, with a large number of examples of receptors coupled to $G_{i/O}$–$G_{q/11}$ (consistent with the increased frequency of this co-couplers[22]).

G_S-$G_{i/O}$ coupling: the GCGR. The GCGR structure has been determined when coupled to G_S (primary coupling) and G_{i1} (secondary coupling). The active GCGR shows a large 19 Å swing of TM6 characteristic of Class B receptors when coupled to G_S,[58] and is also maintained when coupled to G_{i1} (not characteristic in primary $G_{i/O}$ coupling receptors) (Fig. 2A). Hence, the G_{i1} and G_S α5 share a similar cavity, although the G_{i1} α5 engages in less contacts with a smaller amount of buried surface area. The major differences in the receptor between the G_S and G_{i1} coupled structures are found within the ICLs. ICL2 in the G_{i1} complex swings away from the β1-αN losing the extensive interactions made during G_S coupling (Fig. 2D). ICL1 and ICL3 also contribute with interactions to G_{i1} which upon mutation were found to be functionally important for G_{i1} and, to lesser extent, for G_S coupling.[50]

$G_{q/11}$-G_S (and $G_{i/O}$) coupling: the NK$_1$R and CCK$_A$R. These receptors have marked differences in the degree of preference for the $G_{q/11}$ and G_S proteins, NK$_1$R has slight preference (or no preference depending on source) for G_q while CCK$_A$R has up to 1000 times preference for G_q.[53]

Table 1 Summary of GPCR structures coupled to different Gα proteins.

	Receptor	1' Coupling	2' Coupling	Gs/agonist	Gi/o/agonist	Gq/11/agonist	PDB	Resolution (Å)
G_S–$G_{i/O}$	GCGR	G_S	$G_{i/O}$	—	DNG$_{i1}$/Glucagon	—	6LML	3.90
				DNG$_S$/Glucagon	—	—	6LMK	3.70
$G_{q/11}$–G_S ($G_{i/O}$)	CCK$_A$R	$G_{q/11}$	G_S, $G_{i/O}$	DNG$_S$/Cholecystokinin-8	—	—	7MBX	1.95
				—	—	DNG$_{qi}$/Cholecystokinin-8	7EZM	2.90
				—	DNG$_{i1}$/Cholecystokinin-8	—	7EZH	3.20
	NK$_1$R	$G_{q/11}$, G_S	—	—	—	miniG$_{Sq}$70/Substance-P	7P00	2.71
				miniG$_S$399/Substance-P	—	—	7P02	2.87
G_q–$G_{i/O}$	GPR139	$G_{q/11}$	$G_{i/O}$	—	G$_{i1}$/JNJ-63533054	—	7VUG	3.20
				—	—	miniG$_{Sq}$70/JNJ-63533054	7VUH	3.22
	GHSR	$G_{q/11}$	$G_{i/O}$	—	G$_{i1}$/Ghrelin-27	—	7NA7	2.70
				—	miniG$_{O1}$/Ghrelin	—	7W2Z	2.80
				—	—	miniG$_{Sq}$70/Ghrelin-28	7F9Y	2.90
	MRGPRX2	$G_{q/11}$, $G_{i/O}$	—	—	—	miniG$_{qi}$/Cortistatin 14	7S8L	2.45
				—	DNG$_{i1}$/Cortistatin 14	—	7S8M	2.54
	CCK$_B$R	$G_{q/11}$	$G_{i/O}$	—	DNG$_{i2}$/Gastrin-17	—	7F8V	3.30
				—	—	G$_{qi}$/Gastrin-17	7F8W	3.10

DN: dominant negative Gα protein.

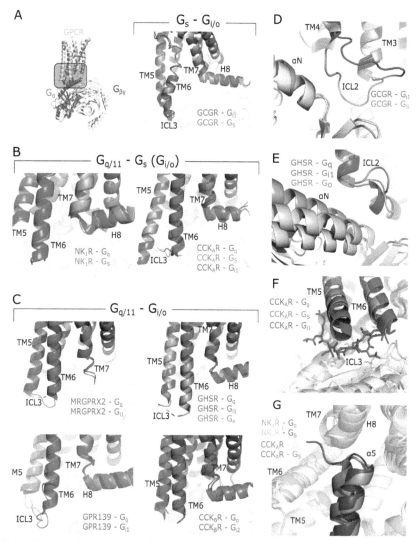

Fig. 2 Structural comparison of receptors coupled to different Gα proteins. Structures aligned on the receptor are shown as blue (G$_{q/11}$), red (G$_i$), orange (G$_O$) and green (G$_S$) cartoons with receptors and G proteins shown in bright and pale colors respectively. Structures of GPCRs bound to different Gα proteins are arranged depending on their co-coupling: G$_S$-G$_{i/O}$ (A); G$_{q/11}$-G$_S$ (and G$_{i/O}$) (B) and G$_{q/11}$–G$_{i/O}$ (C). Conformational changes of receptor ICL2 are shown for GCGR (D) and GSHR (E) and in ICL3 in CCK$_A$R. ICL3 residues are depicted as sticks of their respective colors (F). The differential engagement of the G$_q$ and G$_S$ α5 between the NK$_1$R (blue and red respectively) and CCK$_A$R (yellow and green respectively) is shown in (G) with the Gα and receptors shown as dark and pale colors respectively.

The structures of NK_1R coupled to G_S and G_q show a conserved receptor conformation with a narrow opening of TM6 characteristic of G_q coupling (Fig. 2B). The angle of insertion and rotation of the $G\alpha$ protein relative to the receptor is conserved for both $G\alpha$ proteins and is reminiscent of G_q coupling with extensive interactions between ICL2 and $\beta1$-αN. Overall the NK_1R seems to achieve a similar coupling for both $G\alpha$ proteins by engaging G_S in a G_q-like manner (Fig. 2G). $G_q\alpha5$ binds slightly deeper in the intracellular crevice making just one more interaction than the $\alpha5$ of $G\alpha_S$. The CCK_AR can couple to all four families of $G\alpha$ proteins. CCK_AR structures coupled to G_S, G_q and G_{i1} (structures with $G\alpha$ closest to wild-type were chosen for analysis) show a receptor with small swing of TM6 characteristic of G_q and $G_{i/O}$ coupling (Fig. 2B). In this case, G_q and G_S bind differently (Fig. 2G) with an angle of insertion and rotation that is characteristic of each $G\alpha$ protein (as so does G_{i1}). ICL2 appears more flexible in the G_q and G_i couplings with less interactions at the ICL2-$\beta1$-αN interface than the G_S coupling. In this case, the G_q-like TM6 limits the space available for the bulky G_S and hence, the outmost residues within the "wavy hook" are forced to unwind protruding out of the receptor between TM6 and TM7 (Fig. 2G). ICL3 interacts with G_q and G_i but not with G_S. In this case ICL3 is sandwiched between TM5 and the $G\alpha$ protein and its modification influences specifically the primary G_q coupling[52] (Fig. 2F).

$G_{q/11}$–$G_{i/O}$ couplers: the GSGHR, MRGPRX2, CCK_BR and GPR139: All of the $G_{q/11}$–$G_{i/O}$ coupling receptors are primary couplers to $G_{q/11}$ except for the MRGPRX2 to $G_{i/O}$ is as efficient as to $G_{q/11}$.[57] The angles of insertion and rotation for all $G_{q/11}$ and $G_{i/O}$ couplings are characteristic of each $G\alpha$ protein, and they all display a conserved receptor structure when coupled to $G_{q/11}$ and $G_{i/O}$ (Figs. 2C). In all receptors there is a minor extension/ordering of TM6 when coupled to $G_{i/O}$ in order to keep its conserved interaction with the final aromatic residue in $G_{i/O}$ (G.H5.26, common CGN numbering[5]). ICL2 makes interactions with the $\beta1$-αN in all complexes except for the GSHR coupled to G_{O1} (Fig. 2E). Finally, in the MRGPRX2-G_{i1} complex, ICL3 makes extensive interactions with G_i but it is disordered when coupled to G_q.

5. Insights from structures of GPCRs bound to multiple G proteins

In accordance with previous studies,[45] there is not a simple correlation between selective or promiscuous couplings and sequence conservation.

However, some overall trends arise from these structures of GPCRs coupled to different Gα proteins.

First, promiscuous GPCRs use a similar intracellular cavity for primary and secondary couplings, as determined by the movement (or lack thereof) of the receptor TM6. The outward swing of TM6 is a hallmark of GPCR activation and determines the size and shape of the intracellular cavity. The magnitude of the swing is correlated with the type of Gα coupling (larger for G_S and narrower for $G_{i/O}$ and $G_{q/11}$[38,45,46]). In the available structures of GPCRs coupled to different Gα proteins, TM6 does not change upon coupling to different G proteins, and therefore, primary and secondary Gα proteins are required to use a similar intracellular cavity.

Second, the magnitude of the TM6 outward swing in promiscuous GPCR-G protein pairs is determined by the primary coupler. As an example, the GCGR, uses a wide open TM6 characteristic of its primary G_S coupling, which the secondary G_{i1} is required to use. Conversely, the CCK_AR and NK_1R which are primary G_q couplers adopt a narrower TM6 typical of its G_q coupling, while their secondary G_S is required to adapt to this narrow G_q-like pocket in both of them. It is tempting to speculate that promiscuous GPCRs regulate coupling preference by optimizing the conformation of TM6 to its primary coupler while the secondary coupler will be required to bind "sub-optimally". Of relevance is the unwinding of the "wavy hook" in secondary G_S coupling when bound to the CCK_AR, a feature not present in the NK_1R when bound also to its secondary G_S protein. Such a difference might be the base of their difference in secondary G_S coupling efficacy. In the case of $G_{i/O}$-$G_{q/11}$ co-couplers, all adopt a TM6 conformation that is narrower in comparison to receptors coupling to G_S. The fact that the TM6 outward swing is similar for $G_{i/O}$ and $G_{q/11}$ couplings might explain the fact that receptors that co-couple to $G_{i/O}$ and $G_{q/11}$ are much more abundant than receptors coupling to other pairs.[22] This would be in line with the hypothesis that receptors coupling to G_S are more promiscuous that receptors that couple primarily to $G_{i/O}$,[18,21] however how $G_{q/11}$ primary couplers are more promiscuous than $G_{i/O}$ is unknown.

Third, the angle of insertion and rotation of the Gα protein in comparison to the receptor is normally maintained as is characteristic for each type of Gα protein, with the only exception of the NK_1R-G_S and G_q complexes. This has an impact on the interaction of Gα proteins with the ICLs of the receptors. The fact that the engagement mode is maintained using a different intracellular cavity might support the idea of this interaction to be "sub-optimal."

Finally, the ICLs are the structural elements in the receptors that most change when coupling to different Gα proteins. ICL3 contributes differential interactions between the Gα proteins in 5 out of 7 GPCR-G protein complexes (non in CCK_BR and NK_1R). In the MRGPRX2-G_{i1} and GCGR-G_S there is an ordering of ICL3 to make additional interactions with the Gα protein. In the GSHR-G_O (and not the G_q or G_{i1} complexes), an extension of TM6 contributes additional interactions with the Gα protein and, in the GPR139-G_{i1} the ICL3 rearranges to make a different set of interactions with the Gα protein. Finally, in the CCK_AR there is an increasing ordering of ICL3 to make additional interactions to each Gα protein that correlates with their respective efficacies ($G_q > G_i > G_S$) (consensus efficacy ranking in both Inoue and Avet datasets) (Fig. 2F). Hence, it could be that ICL3 takes a prominent role in regulating G protein coupling efficacy in CCK_AR. There is no correlation between ordering or type of interactions of ICL3 and primary/secondary couplings. Previous studies using chimeric receptors with exchanged ICL3s having a functional impact on G protein promiscuity support the role of ICL3 in promiscuous Gα couplings.[44] However, it seems that there are divergent modes of using ICL3 to regulate Gα protein promiscuity. ICL2 changes conformation or interactions (correlated to the different angle of insertion/rotation of the Gα protein) in most GPCR-G protein complexes. The most prominent conformational changes in ICL2 occur in the GCGR and the GSHR where it forms extensive interactions with the β1-αN in the G_S and G_q complexes respectively and loses all interaction when coupled with G_i and G_O respectively. The interactions between ICL2 and the β1-αN junction are important for G protein selectivity and hence, promiscuity,[59,60] however no patterns can be extracted from the current dataset. Overall, they follow the engagement mode of each type of Gα protein where G_S and $G_{q/11}$ make extensive interactions with β1-αN and $G_{i/O}$ shows weaker interactions. Finally, ICL1 is seen to interact in a functional manner with G_{i1} and not G_S in the GCGR. However, no other receptor shows a differential interaction of ICL1 within their different Gα protein couplings. Based on these structures, care must be taken when using chimeric Gα proteins for structural studies so as not to distort interactions outside the α5 of the Gα protein taking place through the ICLs.

6. Summary and future perspectives

Overall, GPCR-G protein promiscuous couplings occur through the same intracellular cavity whose features seem to be dictated by the primary

Gα coupling, while the ICLs take a prominent role in making differential interactions during promiscuous couplings. There seems to be divergent roles for GPCRs using ICLs in promiscuous couplings, which seem to be specific for each GPCR-G protein pairs (at least with current data). This information could guide drug development, e.g., regulation of promiscuous GPCR-G protein activation through modulation of ICLs. Additional structural information as well as more established GPCR-G protein couplings will aid in the determination of the selectivity barcode and mechanisms of GPCR-G protein promiscuity. A better chance of finding a more defined sequence barcode for GPCR-G protein selectivity might be to search in more segregated groups such as selective GPCRs of a particular type or promiscuous GPCRs with the same primary coupler. However, given the seeming complexity of GPCR selectivity, where promiscuous GPCRs activate differentially, in efficacy and kinetics, different families and sub-types of Gα protein, a unique selectivity barcode for each GPCR and Gα protein set might be possible.

Acknowledgments

We thank Christopher G. Tate for helpful comments on the manuscript.

Author contributions

ACA, JMF, SAU and JGN contributed to all aspects of this chapter.

Conflicts of interest

ACA, JMF, SAU and JGN declare no conflicts of interest.

Funding information

The work in JGN's laboratory is funded by the Ministerio de Ciencia, Innovación y Universidades (PID2020-113359GA-I00), the Spanish Ramón y Cajal program and the FondoEuropeo de Desarrollo Regional (FEDER). SAU is funded by a PhD fellowship of the Diputación General de Aragón (DGA).

References

1. Pierce KL, Premont RT, Lefkowitz RJ. Seven-transmembrane receptors. *Nat Rev Mol Cell Biol.* 2002;3(9):639–650. https://doi.org/10.1038/nrm908.
2. Santos R, Ursu O, Gaulton A, et al. A comprehensive map of molecular drug targets. *Nat Rev Drug Discov.* 2016;16(1):19–34. https://doi.org/10.1038/nrd.2016.230.
3. Sriram K, Insel PA. G protein-coupled receptors as targets for approved drugs: How many targets and how many drugs? *Mol Pharmacol.* 2018;93(4):251–258. https://doi.org/10.1124/mol.117.111062.

4. Rosenbaum DM, Rasmussen SGF, Kobilka BK. The structure and function of G-protein-coupled receptors. *Nature.* 2009;459(7245):356–363. https://doi.org/10. 1038/nature08144.

5. Flock T, Ravarani CNJ, Sun D, et al. Universal allosteric mechanism for Gα activation by GPCRs. *Nature.* 2015;524(7564):173–179. https://doi.org/10.1038/nature14663.

6. Oldham WM, Hamm HE. Heterotrimeric G protein activation by G-protein-coupled receptors. *Nat Rev Mol Cell Biol.* 2008;9(1):60–71. https://doi.org/10.1038/nrm2299.

7. Crilly SE, Ko W, Weinberg ZY, Puthenveedu MA. Conformational specificity of opioid receptors is determined by subcellular location irrespective of agonist. *elife.* 2021;10:1–20. https://doi.org/10.7554/eLife.67478.

8. Irannejad R, Pessino V, Mika D, et al. Functional selectivity of GPCR-directed drug action through location bias. *Nat Publ Gr.* 2017;13(7):799–806. https://doi.org/10. 1038/nchembio.2389.

9. Selma EA, Charlotte K, Isabella M, et al. Receptor-associated independent cAMP nanodomains mediate spatiotemporal specificity of GPCR signaling. *Cell.* 2022;185 (7):1130–1142. https://doi.org/10.1016/j.cell.2022.02.011 Full text linksCite.

10. Polit A, Rysiewicz B, Mystek P, Błasiak E, Dziedzicka-Wasylewska M. The Gαi protein subclass selectivity to the dopamine D2 receptor is also decided by their location at the cell membrane. *Cell Commun Signal.* 2020;18(1):1–16. https://doi.org/10.1186/ s12964-020-00685-9.

11. Uhlén M, Björling E, Agaton C, et al. A human protein atlas for normal and cancer tissues based on antibody proteomics. *Mol Cell Proteomics.* 2005;4(12):1920–1932. https:// doi.org/10.1074/mcp.M500279-MCP200.

12. Flock T, Hauser AS, Lund N, Gloriam DE, Balaji S, Babu MM. Selectivity determinants of GPCR-G protein binding. *Nature.* 2017;545(7654):1–33. https://doi.org/10.1038/ nature22070.

13. Masuho I, Ostrovskaya O, Kramer GM, Jones CD, Xie K, Martemyanov KA. Distinct profiles of functional discrimination among G proteins determine the actions of G protein-coupled receptors. *Sci Signal.* 2015;8(405):1–16. https://doi.org/10.1126/sci-signal.aab4068.

14. Lane JR, May LT, Parton RG, Sexton PM, Christopoulos A. Akinetic view of GPCR allostery and biased agonism. *Nat Chem Biol.* 2017;13(9):929–937. https://doi.org/10. 1038/nchembio.2431.

15. Wootten D, Christopoulos A, Marti-Solano M, Babu MM, Sexton PM. Mechanisms of signalling and biased agonism in G protein-coupled receptors. *Nat Rev Mol Cell Biol.* 2018;19(10):638–653. https://doi.org/10.1038/s41580-018-0049-3.

16. Armstrong JF, Faccenda E, Harding SD, et al. The IUPHAR/BPS guide to PHARMACOLOGY in 2020: Extending immunopharmacology content and introducing the IUPHAR/MMV guide to MALARIA PHARMACOLOGY. *Nucleic Acids Res.* 2020;48(D1):D1006–D1021. https://doi.org/10.1093/nar/gkz951.

17. Sandhu M, Touma AM, Dysthe M, Sadler F, Sivaramakrishnan S, Vaidehi N. Conformational plasticity of the intracellular cavity of GPCR − G-protein complexes leads to G-protein promiscuity and selectivity. *Proc Natl Acad Sci U S A.* 2019;116 (24):11956–11965. https://doi.org/10.1073/pnas.1820944116.

18. Okashah N, Wan Q, Ghosh S, et al. Variable G protein determinants of GPCR coupling selectivity. *Proc Natl Acad Sci U S A.* 2019;116(24):12054–12059. https://doi.org/10. 1073/pnas.1905993116.

19. Wright SC, Bouvier M. Illuminating the complexity of GPCR pathway selectivity—advances in biosensor development. *Curr Opin Struct Biol.* 2021;69:142–149. https:// doi.org/10.1016/j.sbi.2021.04.006.

20. Olsen RHJ, DiBerto JF, English JG, et al. TRUPATH, an open-source biosensor platform for interrogating the GPCR transducerome. *Nat Chem Biol.* 2020. https://doi.org/ 10.1038/s41589-020-0535-8. Published online.

21. Inoue A, Raimondi F, Kadji FMN, et al. Illuminating G-protein-coupling selectivity of GPCRs. *Cell*. 2019;177(7):1933–1947.e25. https://doi.org/10.1016/j.cell.2019.04.044.

22. Hauser AS, Avet C, Normand C, et al. Common coupling map advances GPCR-G protein selectivity. *eLife*. 2022;11:1–31. https://doi.org/10.7554/elife.74107.

23. Avet C, Mancini A, Breton B, et al. Effector membrane translocation biosensors reveal G protein and β arrestin coupling profiles of 100 therapeutically relevant GPCRs. *eLife*. 2022;11:1–34.

24. Kooistra AJ, Mordalski S, Pándy-Szekeres G, et al. GPCRdb in 2021: integrating GPCR sequence, structure and function. *Nucleic Acids Res*. 2021;49(D1):D335–D343. https://doi.org/10.1093/nar/gkaa1080.

25. Pándy-Szekeres G, Esguerra M, Hauser AS, et al. The G protein database, GproteinDb. *Nucleic Acids Res*. 2022;50(D1):D518–D525. https://doi.org/10.1093/nar/gkab852.

26. Jiang M, Bajpayee NS. Molecular mechanisms of Go signaling. *Neurosignals*. 2009;17(1):23–41. https://doi.org/10.1159/000186688.

27. Anderson A, Masuho I, De Velasco EMF, et al. GPCR-dependent biasing of GIRK channel signaling dynamics by RGS6 in mouse sinoatrial nodal cells. *Proc Natl Acad Sci U S A*. 2020;117(25):14522–14531. https://doi.org/10.1073/pnas.2001270117.

28. Rasmussen SGF, Devree BT, Zou Y, et al. Crystal structure of the b 2 adrenergic receptor—Gs protein complex. *Nature*. 2011;2–10. https://doi.org/10.1038/nature10361. Published online.

29. Scheerer P, Park JH, Hildebrand PW, et al. Crystal structure of opsin in its G-protein-interacting conformation. *Nature*. 2008;455(7212):497–502. https://doi.org/10.1038/nature07330.

30. Ching-Ju T, Filip P, Rony N, et al. Crystal structure of rhodopsin in complex with a mini-Go sheds light on the principles of G protein selectivity. *Sci Adv*. 2018;4(9), eaat7052. https://doi.org/10.1126/sciadv.aat7052.

31. Kühlbrandt W. The resolution revolution. *Science (80)*. 2014;343:1443–1444. https://doi.org/10.1126/science.1251652.

32. García-Nafría J, Tate CG. Cryo-electron microscopy: moving beyond X-ray crystal structures for drug receptors and drug development. *Annu Rev Pharmacol Toxicol*. 2020;60(1). https://doi.org/10.1146/annurev-pharmtox-010919-023545.

33. Garcia-Nafria J, Yang L, Xiaochen B, Byron C, Tate CG. Cryo-EM structure of the adenosine A2A receptor coupled to an engineered heterotrimeric G protein. *eLife*. 2018;7:1–19. e35946.

34. García-Nafría J, Nehmé R, Edwards PC, Tate CG. Cryo-EM structure of the serotonin 5-HT1B receptor coupled to heterotrimeric Go. *Nature*. 2018;558(7711):620–623. https://doi.org/10.1038/s41586-018-0241-9.

35. Draper-Joyce CJ, Khoshouei M, Thal DM, et al. Structure of the adenosine-bound human adenosine A1 receptor–Gi complex. *Nature*. 2018;558(7711):559–563. https://doi.org/10.1038/s41586-018-0236-6.

36. Koehl A, Hu H, Maeda S, et al. Structure of the μ opioid receptor-Gi protein complex. *Nature*. 2018;1–23. https://doi.org/10.1038/s41586-018-0219-7. Published online.

37. Kang Y, Kuybeda O, De Waal PW, et al. Cryo-EM structure of human rhodopsin bound to an inhibitory G protein. *Nature*. 2018;558(7711):553–558. https://doi.org/10.1038/s41586-018-0215-y.

38. Maeda S, Qu Q, Robertson MJ, Skiniotis G, Kobilka BK. Structures of the M1 and M2 muscarinic acetylcholine receptor/G-protein complexes. *Science (80)*. 2019;364(6440):552–557. https://doi.org/10.1126/science.aaw5188.

39. García-Nafría J, Tate CG. Structure determination of GPCRs: Cryo-EM compared with X-ray crystallography. *Biochem Soc Trans*. 2021;49(5):2345–2355. https://doi.org/10.1042/BST20210431.

40. Berman HM, Westbrook J, Feng Z, et al. The protein data bank. *Nucleic Acids Res.* 2000;28:235–242. http://www.rcsb.org/pdb/status.html.
41. Venkatakrishnan AJ, Deupi X, Lebon G, et al. Diverse activation pathways in class a GPCRs converge near the G-protein-coupling region. *Nature.* 2016;536(7617): 484–487. https://doi.org/10.1038/nature19107.
42. Zhou Q, Yang D, Wu M, et al. Common activation mechanism of class a GPCRs. *elife.* 2019;8:1–31. https://doi.org/10.7554/eLife.50279.
43. Conklin BR, Farfel Z, Lustig KD, Julius D, Bourne HR. Substitution of three amino acids switches receptor specificity of Gqα to that of Giα. *Nature.* 1993;363:274–276.
44. Wong SK-F, Ross EM. Chimeric muscarinic cholinergic-adrenergic receptors that are functionally promiscuous among G proteins. *J Biol Chem.* 1994;269:18968–18976.
45. García-Nafría J, Tate CG. Cryo-EM structures of GPCRs coupled to Gs, Gi and Go. *Mol Cell Endocrinol.* 2019;(488):1–13. https://doi.org/10.1016/j.mce.2019.02.006.
46. Glukhova A, Draper-Joyce CJ, Sunahara RK, Christopoulos A, Wootten D, Sexton PM. Rules of engagement: GPCRs and G proteins. *ACS Pharmacol Transl Sci.* 2018;1 (2):73–83. https://doi.org/10.1021/acsptsci.8b00026.
47. Yang F, Mao C, Guo L, et al. Structural basis of GPBAR activation and bile acid recognition. *Nature.* 2020;587(7834):499–504. https://doi.org/10.1038/s41586-020-2569-1.
48. Nojima S, Fujita Y, Kimura KT, et al. Cryo-EM structure of the prostaglandin E receptor EP4 coupled to G protein. *Structure.* 2021;29(3):252–260.e6. https://doi.org/10.1016/j.str.2020.11.007.
49. Qin J, Cai Y, Xu Z, et al. Molecular mechanism of agonism and inverse agonism in ghrelin receptor. *Nat Commun.* 2022;13(1). https://doi.org/10.1038/s41467-022-27975-9.
50. Qiao A, Han S, Li X, et al. Structural basis of Gs and Gi recognition by the human glucagon receptor. *Science.* 2020;1352:1346–1352.
51. Thom C, Ehrenmann J, Vacca S, et al. Structures of neurokinin 1 receptor in complex with Gq and Gs proteins reveal substance P binding mode and unique activation features. *Sci Adv.* 2021;7(50):30–32. https://doi.org/10.1126/sciadv.abk2872.
52. Liu Q, Yang D, Zhuang Y, et al. Ligand recognition and G-protein coupling selectivity of cholecystokinin a receptor. *Nat Chem Biol.* 2021;17(12):1238–1244. https://doi.org/10.1038/s41589-021-00841-3.
53. Mobbs JI, Belousoff MJ, Harikumar KG, et al. Structures of the human cholecystokinin 1 (CCK1) receptor bound to Gs and Gq mimetic proteins provide insight into mechanisms of G protein selectivity. *PLoS Biol.* 2021;19(6):1–28. https://doi.org/10.1371/journal.pbio.3001295.
54. Wang Y, Guo S, Zhuang Y, et al. Molecular recognition of an acyl-peptide hormone and activation of ghrelin receptor. *Nat Commun.* 2021;12(1):1–9. https://doi.org/10.1038/s41467-021-25364-2.
55. Zhang X, He C, Wang M, et al. Structures of the human cholecystokinin receptors bound to agonists and antagonists. *Nat Chem Biol.* 2021;17(12):1230–1237. https://doi.org/10.1038/s41589-021-00866-8.
56. Zhou Y, Daver H, Trapkov B, et al. Molecular insights into ligand recognition and G protein coupling of the neuromodulatory orphan receptor GPR139. *Cell Res.* 2021;1–4. https://doi.org/10.1038/s41422-021-00591-w. Published online.
57. Cao C, Kang HJ, Singh I, et al. Structure, function and pharmacology of human itch GPCRs. *Nature.* 2021;600(7887):170–175. https://doi.org/10.1038/s41586-021-04126-6.
58. Dal Maso E, Glukhova A, Zhu Y, et al. The molecular control of calcitonin receptor signaling. *ACS Pharmacol Transl Sci.* 2019;2(1):31–51. https://doi.org/10.1021/acsptsci.8b00056.

59. Jelinek V, Mösslein N, Bünemann M. Structures in G proteins important for subtype selective receptor binding and subsequent activation. *Commun Biol.* 2021;4 (1). https://doi.org/10.1038/s42003-021-02143-9.

60. Duan J, Shen D-D, Zhao T, et al. Molecular basis for allosteric agonism and G protein subtype selectivity of galanin receptors. *Nat Commun.* 2022;13(1):1–13. https://doi.org/10.1038/s41467-022-29072-3.

CHAPTER NINE

Opioid signaling and design of analgesics

Barnali Paul, Sashrik Sribhashyam, and Susruta Majumdar*

Center for Clinical Pharmacology, University of Health Sciences & Pharmacy at St Louis and Washington University School of Medicine, St Louis, MO, United States
*Corresponding author: e-mail address: susrutam@email.wustl.edu

Contents

Abstract

Clinical treatment of acute to severe pain relies on the use of opioids. While their potency is significant, there are considerable side effects that can negatively affect patients. Their rise in usage has correlated with the current opioid epidemic in the United States, which has led to more than 70,000 deaths per year (Volkow and Blanco, 2021). Opioid-related drug development aims to make target compounds that show strong potency but with diminished side effects. Research into pharmaceuticals that could act as potential alternatives to current pains medications has relied on mechanistic insights of opioid receptors, a class of G-protein coupled receptors (GPCRs), and biased agonism, a common phenomenon among pharmaceutical compounds where downstream effects can be altered at the same receptor via different agonists. Opioids function typically by binding to an active site on the extracellular portion of opioid receptors. Once activated, the opioid receptor initiates a G-protein signaling pathway and/or the β-arrestin2 pathway. The proposed concept for the development of safe analgesics around mu and kappa opioid receptor subtypes has focused on not recruiting β-arrestin2 (biased agonism) and/or having low efficacy at the receptor (partial agonism). By altering chemical motifs on a

Progress in Molecular Biology and Translational Science, Volume 195
ISSN 1877-1173
https://doi.org/10.1016/bs.pmbts.2022.06.017
153

common scaffold, chemists can take advantage of biased agonism as well as create compounds with low intrinsic efficacy for the desired treatments. This review will focus on ligands with bias profile, signaling aspects of the receptor and probe into the structural basis of receptor that leads to bias and/or partial agonism.

Abbreviations

β FNA	beta-funaltrexamine
BRET	bioluminescence resonance energy transfer
cAMP	cyclic adenosine monophosphate
CPA	conditioned place aversion
CPP	conditioned place preference
CryoEM	cryogenic electron microscopy
DAMGO	[DAla2, N-MePhe4, Gly-ol5]-enkephalin
DOR	delta opioid receptor
FDA	food and drug administration
GDP	guanosine diphosphate
Gpcrs	G-protein coupled receptors
GRK2	G-protein coupled receptor kinase 2
GTP	guanosine triphosphate
HTS	high-throughput screening
KOR	kappa opioid receptor
MAPK	mitogen-activated protein kinase
MOR	mu-opioid receptors
mTOR	mammalian target of rapamycin
Nb33 assay	nanobody-based biosensor assay
NOP	nonclassical nociceptin opioid receptor
NSAIDS	non-steroidal anti-inflammatory drugs
OUD	opioid use disorder
PI3K	phosphoinosite 3 kinase
SAR	structure activity relationship
Sc	subcutaneous
SSRI's	selective serotonin reuptake inhibitors
VTA	ventral tegmental area

1. Introduction

Pain is classified as acute (severe and short-lived) and chronic (mild to severe with prolonged effect). Four major classes of analgesics are typically used for the treatment of pain, these include anti-inflammatory agents, antidepressants like selective serotonin reuptake inhibitors (SSRIs), anti-convulsants like gabapentinoids, and finally opioids.[1] Non-steroidal anti-inflammatory drugs (NSAIDs), such as aspirin and paracetamol (acetaminophen) treat pain by reducing inflammation. Despite the extensive therapeutic utility, NSAIDs are notorious for severe side effects such as gastrointestinal toxicities, cardiovascular complications, renal injuries, hepatotoxicity, and hypertension, whereas, acetaminophen is critically noxious to the liver.[2-4] SSRIs are notable candidates for the treatment of pain. Although they have a considerably better adverse effect profile, their application is typically limited due to the lack of efficacy. Gabapentin, an anticonvulsant, has been used to treat chronic pain. Nevertheless, due to the lack of uniformity in efficacy,[5,6] it often fails to provide sufficient analgesia in patients. Analgesic opioids (narcotics) operate by changing the brain's perception of pain. Morphine, codeine, fentanyl, hydrocodone, methadone, and oxycodone are commonly used opioids for the treatment of acute pain. Opioids are very efficient drugs, but their long-term use to treat chronic pain can cause life-threatening side effects. In the last two decades, prescriptions of opioids have increased the risk of a fatal overdose by around four times.[7]

Opioid receptors are one of the most important targets in drug design.[8] However, the lack of proper understanding of their physiology and mechanism of action provides an opportunity to find out mysteries in opioid receptor structure and its function to develop novel pain-relieving therapeutics with attenuated side effects.[9] These receptors belong to the group of seven transmembrane domain G protein-coupled receptors (GPCRs). Four main subtypes of opioid receptors are reported, and they are referred to as the μ-opioid receptors (MOR), δ-opioid receptors (DOR), κ-opioid receptors (KOR), and the nonclassical nociceptin opioid receptor (NOP).[10-12] GPCRs are present at the cellular membrane with an orthosteric binding site on the cell's surface. The signaling mechanism begins with binding to the orthosteric site which causes conformational changes along the intracellular portion of the receptor (Fig. 1A). Activation of the receptor allows for the recruitment of the heterotrimeric G-protein, which is comprised of 3 separate units, Gα, Gβ, and Gγ. All four types of opioid receptors can be activated by exogenous and endogenous opioids. This process stimulates GDP dissociation (the inactive heterotrimeric G-protein allows GDP to bind at Gα) and subsequent GTP attachment to Gα. This signaling sequence results in noticeable conformation changes to the G-protein unit. Detachment of both subunits promotes different downstream signaling process. Inhibitory Gα involves a reduction in cyclic adenosine monophosphate (cAMP) production by attenuating the activity of effector protein kinases (cAMP-dependent protein kinase) and reducing neuronal discharge. The Gβγ subunit stops the signal transduction by directly interacting with the ion channels and suppressing the Ca^{2+} influx. This causes a decrease in the impulsiveness of neurons and thus inhibiting the release of pronociceptive neuropeptides.

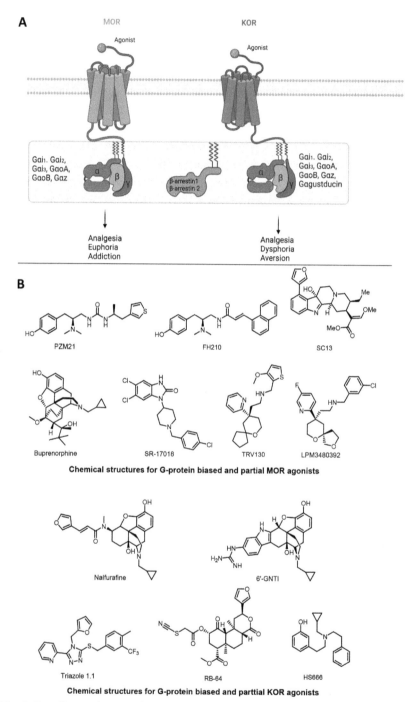

Fig. 1 Signaling pathway and reported chemical structures of MOR and KOR agonists. (A) MOR and KOR activation through external agonists and corresponding downstream signaling produces analgesia as well as adverse effects. (B) Reported small molecule chemical ligands behaving as biased and/or partial agonists at MOR and KOR.

Gβγ subunit regulates G protein-coupled phosphoinosite 3 kinase (PI3K) and/or mitogen-activated protein kinase (MAPK) pathways. Along with this "classical" G-protein pathway, another downstream signaling route has been reported which involves β-arrestins2 (Fig. 1A).[13,14] Upon GPCR activation, MOR and KOR undergo phosphorylation by G protein-coupled receptor kinase 2 (GRK2) leading to β-arrestin2 recruitment. This process regulates opioid receptor signaling through desensitization and clathrin-mediated internalization.[14] Several structural studies reveal that receptors can preferentially interact and induce specific conformational changes by activating individual signaling pathways either via G-proteins or β-arrestin2. [15,16]

The reported concept provides valuable insights regarding the selective binding mode of opioids involving specific residues on receptors, although more studies are required to fully understand the mechanism of action. The preferential stimulation of a certain signaling pathway is called biased agonism or functional selectivity.[17–19] However, recent studies indicate several ambiguities in the biased agonism hypothesis both at MOR as well as at KOR.[18,20–23] In contrast to previous reports, some adverse effects like respiratory depression remain intact in absence of the β-arrestin2 activation pathway.[22,24] Herein, the concept of low efficacy partial agonism introduces a suitable correlation between therapeutic activity and efficacy without bias factors.[25] The concept of functional selectivity along with partial agonism opens a new era for the development of safer opioid analgesics with fewer adverse effects while retaining efficacy. Herein, we will discuss the pharmacology of a few reported MOR and KOR agonists (Fig. 1B) and future directions to develop better drug candidates.

2. Mu opioid receptor

Opioids are clinically used for the treatment of pain, although their usage is not recommended for the treatment of central neuropathic pain anymore.[1] In the past few decades, the major problem with conventionally prescribed opioids is that at high doses they cause severe adverse effects, and at an overdose, they can be lethal.[8,26] Functional selectivity or biased agonism has been proposed as a phenomenon aimed at differentiating between the specific receptor activations (G-proteins over β-arrestins) followed by distinguishable downstream signaling pathways when activated by opioids to develop safer analgesia with reduced side effects.[27,28] At MOR it has been reported that opioid-related analgesia originates from G-protein dependent pathway whereas tolerance, addiction, and respiratory depression are dependent on the β-arrestin pathway.[28,29] These initial findings signify a preferential activation and signal transduction by G-protein-biased agonists over β-arrestin2 to diminish side effects. Phosphorylation-deficient MOR mutant mice that would not recruit β-arrestin2 exhibit promising analgesia but with induced respiratory depression, constipation, and hyperlocomotion. Along with this, morphine and fentanyl-induced respiratory depression is also preserved in β-arrestin2 knockout mice.[22,24] This contemporary knowledge suggests that the concept of bias agonism alone may not be enough to provide a probable explanation for reducing the side effects of MOR-mediated analgesia. Recent studies by Gillis et al.[30] highlight that the G-protein-biased agonists (TRV130, PZM21, and SR-17018) show lower intrinsic efficacies towards the receptor compared to agonists with higher intrinsic efficacies (fentanyl and morphine). In addition, the low intrinsic efficacy MOR agonists also show far reduced respiratory depression compared to agonists with higher intrinsic efficacy at the receptor. Hence, the concept of partial agonism suggests a novel mechanism with an advanced therapeutic profile for new opioid ligand development.[25,30–34] The next section covers the pharmacology of biased and partial agonists (see Fig. 1B for the chemical structures of these agents).

2.1 Putative G-protein biased ligands

The concept of G-protein-biased agonism creates new pillars to build analgesia with an advanced safety profile targeting MOR. TRV130 or Oliceridine is an FDA-approved drug, which has been developed from a high throughput screening (HTS) platform followed by chemical optimization to improve its pharmacology and selectivity towards MOR.[33–36] Since TRV130 is moderately effective in separating respiratory depression from analgesia in humans and other adverse effects in rodents, it allows for testing the biased agonism hypothesis in humans. PZM21 similarly was discovered through a structure-based platform.[35] It was proposed to be a G-protein-biased agonist with decreased respiratory depression.[30,35] In rodents, its rewarding properties were limited,[35,36] though intrathecally administered PZM21 still retained abuse liability.[37] SR17018, a substituted piperidine benzimidazole-based analog, showed a high agonistic affinity for the MOR with no respiratory suppression.[27,30] Several reviews have covered the pharmacology of these potent candidates in detail.[8,18]

Recently a spirocyclic analog, LPM3480392 has been reported as a MOR selective biased agonist. This compound contains a structural modification of the marketed therapeutic TRV130 and it is a G-protein biased agonist at MOR like TRV130.[38,39] Although it functions as a partial MOR agonist like TRV130, LPM3480392 exhibits significant MOR agonistic activity (cAMP, $EC_{50} = 0.35$ nM, $E_{max} = 91\%$) without β-arrestin2 recruitment ($EC_{50} > 30,000$ nM, $E_{max} = 2\%$). In comparison, TRV130 in the same assay had cAMP, $EC_{50} = 1.8$ nM, $E_{max} = 92\%$ and β-arrestin2 $EC_{50} = 12$ nM, $E_{max} = 5\%$). Along with this, LPM3480392 showed good brain penetration with a promising pharmacokinetic profile and produced a potent antinociceptive effect with reduced respiratory suppression as compared to morphine and TRV130. LPM3480392 did not show significant respiratory depression in mice at a dose of 0.3 mg/kg (SC) which is only 1.5 times greater than its antinociceptive ED_{50} value. It showed antinociception at a lower dose (0.3 mg/kg) compared to morphine (7 mg/kg). LPM3480392, a partial agonist of MOR, is currently undergoing Phase I clinical trials (CTR20210370) as a potential pain-relieving candidate.[40]

2.2 MOR partial agonists

While first-generation G-protein biased agonists have provided avenues to separate respiratory depression from analgesia. The concept has also led to scientists revisiting the concept of partial agonism.[21,34] The next section discusses two ligands, an established partial agonist and another new ligand that appears promising in preclinical models.

2.2.1 Buprenorphine

Buprenorphine (BUP), first synthesized and evaluated over 40 years ago, is the FDA-approved drug for the treatment of opioid use disorder (OUD).[41] It is well characterized to blunt the adverse effects of many opioids, such as heroin. However, depending on the type of receptor buprenorphine shows multifunctional behavior. It acts as a partial agonist for MOR showing no β-arrestin recruitment,[42] is an antagonist at DOR[42] and KOR[42], and a weak agonist at NOP[43]. Buprenorphine is a potent analgesic[44] in rodents and humans but is known to show a ceiling effect in respiratory depression assays.[45] However, it is not free from other side effects, as it shows hyperlocomotion, constipation, and reward-like behavior in rodents.[42,46] The application of BUP is further limited by its active metabolites norbuprenorphine (norBUP), norbuprenorphine-3-glucuronide, and buprenorphine 3-glucuronide. The lower potency but higher efficacy of norBUP at MOR and unique biological activity of other glucuronides contribute strongly to the overall pharmacology of buprenorphine and thus making these metabolites[47–50] critical substrates in the overall pharmacology of buprenorphine. While buprenorphine is no doubt a safer analgesic, its metabolic liability and activity at opioid and NOP[51] complicates its pharmacology. A metabolically stable buprenorphine-like analog is desirable to ideally test the low efficacy model.

2.2.2 SC13

Mitragyna speciosa plant (commonly known as Kratom) is enriched with biologically privileged natural products for the treatment of pain because of their remarkable opioid properties and stimulant-like effects.[52,53] Mitragynine, one of the major indole alkaloids isolated from kratom, along with semi-synthetic natural products, 7-OH mitragynine, and mitragynine pseudoindoxyl were reported as G-protein-biased agonists showing suitable analgesic properties.[53–61] As a follow-up study, Chakraborty et al. [25] have developed a structure-activity relationship (SAR) based on a new series of C9 derivatized 7-OH mitragynine analogs. Their research work came up with a suitable drug-like candidate, SC13, which has partial agonism (Fig. 2) with lower efficacy than DAMGO or morphine in heterologous G-protein assays (Fig. 2A) and synaptic physiology (Fig. 2D). SC13 has less efficacy relative to other potent MOR agonists such as DAMGO, fentanyl, and morphine (Fig. 2A) but lower efficacy than buprenorphine in assays with limited MOR reserve. In the BRET-based Nb33 recruitment assay[30] (an assay with limited receptor reserve discussed in detail in later sections) SC13 has an $E_{max} = 21\%$ and 8% against hMOR (Fig. 2C) and mMOR respectively, which is completely relatable with the reported efficacy at MOR of synthetic small molecule partial agonists such as TRV130 (42%)[30], SR17018 (20%)[30], PZM21 (38%)[30] and buprenorphine (24%)[30]; furthermore, SC13 has much lower efficacy than the full agonist morphine (72%)[30]. SC13 also showed poor arrestin recruitment across a battery of assays for example TANGO ($E_{max} = 45\%$), Path hunter Discoverx ($E_{max}\% = <20\%$) and BRET (bioluminescence resonance energy transfer) assays based βarrestin1 ($E_{max} = <10\%$) and βarrestin2 ($E_{max} = <10\%$, Fig. 2B) when normalized against DAMGO. The authors also performed the BRET-based TRUPATH assay[62] to study the activity of each of the Gα-subtypes. They determined that SC13 showed lower efficacy than DAMGO, morphine, or fentanyl at each Gi/o/z subtype and comparable intrinsic efficacy with buprenorphine at GoA, GoB, and Gz. This result attracts enormous interest as the most abundant Gα subunit in the brain is GoA (at GoA SC13: $E_{max} = 75\%$; Buprenorphine: $E_{max} = 65\%$).

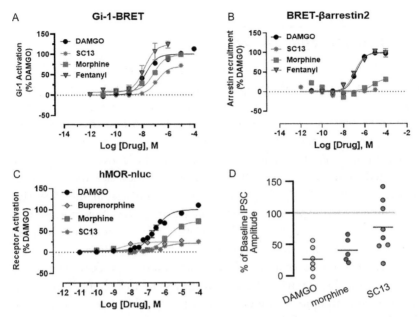

Fig. 2 G-protein signaling, arrestin signaling, whole cell electrophysiology of MOR ligands. (A) SC13 (blue) is MOR partial agonist with lower efficacy compared to morphine (purple), fentanyl (brown), and DAMGO (black) in Gi-1 BRET assay. (B) SC13 (blue) shows no β-arrestin2 recruitment compared to fentanyl (brown) and DAMGO (black) in BRET arrestin recruitment assay. (C) SC13 shows lower efficacy compared to DAMGO and morphine and similar efficacy to buprenorphine in Nb33 recruitment assays measured using BRET assays in hMOR. (D) Inhibition of electrically evoked IPSCs in VTA neurons in response to DAMGO (5 μM), morphine (10 μM) and SC13 (10 μM), where each circle is one neuron and horizontal bars indicate means. *Reproduced with permission from Chakraborty S, DiBerto JF, Faouzi A, et al. A novel mitragynine analog with low-efficacy Mu opioid receptor agonism displays antinociception with attenuated adverse effects.* J Med Chem. *2021;64(18):13873–13892.*

This advantageous MOR selective partial agonist demonstrated MOR-dependent antinociception and analgesic activity comparable to morphine while lacking conditioned place preference, constipation, respiratory depression, and hyperlocomotion at equianalgesic morphine doses. Moreover, SC13 was reported to be orally active as well as very potent when administrated subcutaneously under tested conditions in mice. This low-efficacy partial agonism hypothesis on the mitragynine template gives a novel approach to lead MOR agonists over buprenorphine with reduced side effects and preferable analgesia.[27] The limitations of SC13 include the presence of a metabolically labile furan ring which second-generation molecules may potentially aim to address.

3. Kappa opioid receptor

The KOR represents a promising target for pain attenuation. Because of its distribution across the peripheral and central nervous system, it attracts attention in a variety of novel areas including cardiovascular disease, pruritus, nausea, inflammatory diseases, spinal anesthesia, stroke, hypoxic pulmonary hypertension, multiple sclerosis, addiction, and post-traumatic cartilage degeneration.[63] KOR agonists do not activate the conventional pathway stimulated by morphine-like MOR agonists and they lack the side effects associated with MOR agonists, such as respiratory depression; therefore, they are considered to be promising non-addictive analgesics. Nevertheless, their use as therapeutics is inadequate due to other side effects such as sedation, motor incoordination, hallucinations, anxiety, and dysphoria-like states.[64] Understanding the mechanism of KOR activation to produce dysphoria is the key to the development of better analgesics and to defining how endogenous dynorphin opioids produce their depression-like effects. In addition, the aversive effects of KOR activation require arrestin-dependent p38 MAPK stimulation in dopamine neurons but do not require inhibition of dopamine release in the nucleus accumbens. This supports the hypothesis that functionally selective G-protein-biased KOR agonists might have a non-dysphoric, antipruritic, analgesic therapeutic value with low abuse potential.[65,66] Several laboratories have suggested that G protein–biased KOR agonists might be analgesic with fewer side effects. The next section will cover the pharmacology of biased KOR agonists and partial agonists which do not recruit βarrestin-2 (Fig. 1B).

3.1 Putative KOR biased agonists

Quite a few G protein-biased KOR full agonists have been synthesized in recent years. Triazole 1.1 was selected from the high-throughput screening (HTS) of nearly 300,000 compounds and identified as a G–protein-biased KOR agonist in cell lines.[67] Triazole 1.1 retained efficacy similar to conventional KOR agonists in antinociception and antipruritic assays but did not cause sedation and dysphoria/aversion at analgesic doses or higher compared to the balanced KOR agonist U50,488H.[68,69] Whereas, nalfurafine, another G-protein biased agonist[70], produced noteworthy antinociceptive and anti-scratch effects without causing dysphoria, psychotomimesis, sedation, motor incoordination, and conditioned place aversion (CPA) in the effective dose ranges.[71,72] Nalfurafine is also the only clinically used biased KOR agonist. This candidate is used in Japan for the treatment of hemodialysis and chronic liver diseases and is a moderately selective biased KOR agonist and partial MOR agonist.[73,74] Interestingly, RB-64[23] (Artasalvin or 22-thiocyanatosalvinorin A), a semi-synthetic derivative of centrally active salvinorin A (balanced KOR agonist), is G-protein biased KOR agonist and shows significant and long-lasting antinociception. RB-64, however, does induce CPA unlike triazole 1.1 and nalfurafine but lacks other KOR adverse effects like sedation and anhedonia-like actions. In summary, the results of KOR-biased full agonists are like MOR-biased agonists, mixed. It is possible that the promising results with nalfurafine are because of other mechanisms beyond bias (not activating the mTOR pathway for example[71]). The role of amplified assays in assessing the bias of the KOR full agonists needs to be examined more closely (discussed in sections later in this review). As additional reading on this topic, several reviews have covered the pharmacology of these KOR G-protein biased full agonists[18,75] in detail.

3.2 KOR partial agonists

KOR partial agonists have a potential advantage as antidepressants and anxiolytics over a KOR full agonist. Two ligands with KOR partial agonism that display reduced β-arrestin2 recruitment are discussed in this section.

3.2.1 6'GNTI

6'-Guanidinonaltrindole (6'-GNTI) is reported to be a potent partial agonist at KOR for the G protein activation without recruiting β-arrestin2. It is a derivative of the highly potent DOR-selective antagonist Naltrindole (NTI) and the 6'-guanidine side chain is one of the major structural requirements to be selective at KOR over other opioid receptors. It also acts as an antagonist which blocks the β-arrestin2 pathway and KOR internalization stimulated by other unbiased KOR agonists (such as U50,488).[76,77] In endogenous striatal neurons as well the preservation of bias activity for G protein over β-arrestin2-mediated signaling induced by 6'-GNTI actions at KOR was observed.[78] Recent research indicates that 6'-GNTI is a potent anticonvulsant and an antiseizure candidate with reduced dysphoria-like side effects. The lack of CPA or motor impairment *in vivo* makes it a suitable, safer analgesic candidate over other balanced KOR agonists.[79] Among liabilities, it has poor brain penetration, and it is possible that, like MOR-biased agonists, some of its preferable properties arise because of its lower intrinsic efficacy.

3.2.2 HS666

HS666 is a chemical modification of Dopamine D2 receptor agonist RU24213, which is a moderately potent KOR antagonist.[80] HS666 ($EC_{50} = 449$ nM and $E_{max} = 24\%$) showed a selective KOR partial agonism for β-arrestin2 recruitment when compared with U69,593 ($EC_{50} = 68$ nM and $E_{max} = 100\%$) in DiscoveRx PathHunter β-arrestin 2 assay. In GTPγS assay, HS666 showed agonistic properties with G-protein: $EC_{50} = 36$ nM and $E_{max} = 50\%$ in comparison, with the balanced agonist, U69,693 (EC_{50} 18 nM and $E_{max} = 100\%$).[81] Remarkably, HS666 proved to have very promising effects on KOR-specific analgesia, when administered intra-cerebroventricularly, without showing motor impairment and CPP/CPA (Conditioned Place Preference/Conditioned Place Aversion).[81] A few structural modifications have been performed on this scaffold by introducing bulky substitutes as well as enhancing hydrophobicity to develop very potent KOR selective analogs which exhibited analgesia while lacking sedation and motor impairment.[82] In a follow-up study, HS666 was shown to not activate the mTOR pathway. Phosphoproteomic approaches in mouse brains have proposed that KOR-mediated CPA is dependent on this pathway.[71] CPA, induced by balanced KOR agonists, is for example blocked by mTOR inhibitors like rapamycin. The lack of mTOR activation is also seen with another biased KOR agonist nalfurafine which also doesn't cause CPA in rodents and lacks dysphoria in humans. More mechanistic insight is required to understand if the lack of CPA in rodents with HS666 is due to its partial agonistic property or because it does not activate the mTOR pathway. It must be noted that a close analog of HS666 namely HS665[81] does show CPA in rodents possibly owing to its full agonism at KOR.

4. Structural basis of G-protein bias

While numerous MOR and KOR G-protein agonists do exist, the structural basis of receptor-mediated G-protein and arrestin signaling isn't well understood. CryoEM structures of G-protein biased MOR ligands are now available[83] and previous work from Majumdar, Katritch, and Roth groups[32] provide insights into sub-pockets that may drive arrestin dis-engagement. We will cover the structural basis of biased signaling using these concepts in this section.

4.1 Structure based design of MOR biased agonists

A ligand named PZM21 was discovered by computational docking at MOR followed by optimization.[35] As discussed previously, PZM21 was reported to be a G-protein biased partial agonist which showed analgesic actions with attenuated respiratory depression compared to the prototypic MOR agonist morphine.[30,35] A recent structure-based study by Georgios Skiniotis, Brian K. Kobilka, and Peter Gmeiner groups demonstrated that the binding specificity of biased ligands towards G-protein over arrestin recruitment is dependent on lipophilic vestibules formed by TM2, TM3, and ECL1 in MOR. The cryoEM structure of PZM21 (PZM21-µOR-Gi protein complex structure at 2.9 Å resolution) was recently solved by this group.[83] Further optimization and suitable substitution with lipophilic functionalities on the PZM21 scaffold led to an increase in G-protein bias compared to PZM21. The newly developed potential PZM21 analogs have higher activity with the extracellular vestibule, especially with V143$^{3.28}$, I144$^{3.29}$, W133^{ECL1}, and N127$^{2.63}$, and they bestow an increased G protein-biased activity in comparison with unbiased agonists such as DAMGO and BU72. The authors also determined another high-resolution cryoEM structure of FH210 (FH210-MOR-Gi protein complex at a global resolution of 3.0 Å), a naphthyl-substituted acryl amide analog of PZM21. FH210 retained PZM21-like binding modes such as the ionic interaction between the amine group of the ligand and D147$^{3.32}$ and a conventional hydrogen bond with Y326$^{7.43}$. Along with these conventional binding modes, FH210 showed increased Van der Waals interaction within the lipophilic pocket with D216^{ECL2}, C217^{ECL2}, W133^{ECL1} and N127$^{2.63}$. In comparison to the thiophenyl moiety in PZM21; FH210 with a naphthyl moiety showed greater contact ~about 31 Å^2 higher within the lipophilic vestibule formed by TM2, TM3, ECL1, and ECL2. These recent findings provide a novel structure-based evolution approach for the development of potential therapeutic leads with functionally selective agonistic activity and possibly diminished side effects for MOR drug discovery.

4.2 TM5-ECL2 region as a subpocket for morphinan bias

The active state kappa-opioid receptor structure was solved by the Roth group in 2018[84] using a ligand named **MP1104**[85] based on the morphinan template. **MP1104** is a mixed agonist at both MOR as well as KOR and robustly recruited β-arrestin2 at both subtypes. The 6β-amidophenyl arm of this compound was found to bind in the TM2-TM3 region of both MOR as well as KOR. The cyclohexene ring which is a part of ring C of the template prefers a boat conformation. Efforts were next made to design G-protein biased agonists on this template using the **MP1104** bound KOR structure.[32] Saturation of the cyclohexene ring C of the morphinan template led to the synthesis of **MP1202** (Fig. 3Aa), an agonist which now showed G-protein bias at MOR but not KOR (Fig. 3Abc). Computational studies showed the 6β-amidophenyl arm in KOR binds

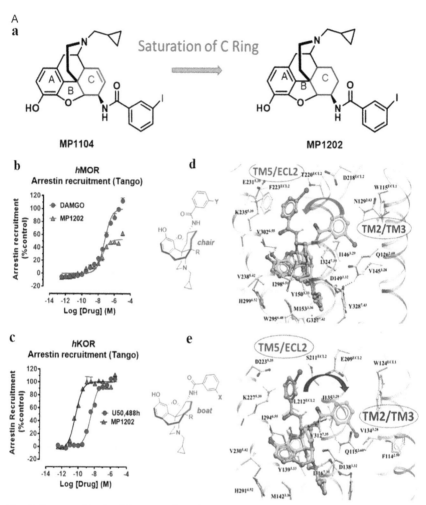

Fig. 3 Structural basis of G-protein bias. (A) MP1202 at KOR targets the TM2-TM3 region while at MOR targets the TM5-ECL2 region and shows distinct signaling properties (a) Chemical structure of MP1104 and 1202. (b) MP1202 (red) is partial agonist in Tango-arrestin recruitment assays compared to DAMGO (blue) at MOR. (c) MP1202 (green) is a full agonist in Tango-arrestin recruitment assays compared to U50,488h (purple) at KOR. (d) The docking poses of MP1202 (chair form, brown stick) and (boat form, green stick) at an active state of MOR, the saturated ring C in MP1202 leads to interaction of the ligand in the ECL2 and TM5 region leading to a preference of chair form shown by a red arrow. (e) The docking poses of MP1202 (chair form, brown stick) and (boat form, green stick) at an active state of KOR, the flip of ring C conformation from chair to boat is shown by a blue arrow. (B) MP1207 and MP1208 prefer the chair conformation and target the TM5-ECL2 region and are G protein biased agonists at KOR and at MOR show no measurable arrestin recruitment. (a) Chemical modification on MP1202. (b) MP1207 (orange) and MP1208 (green) are partial agonists in Tango-arrestin

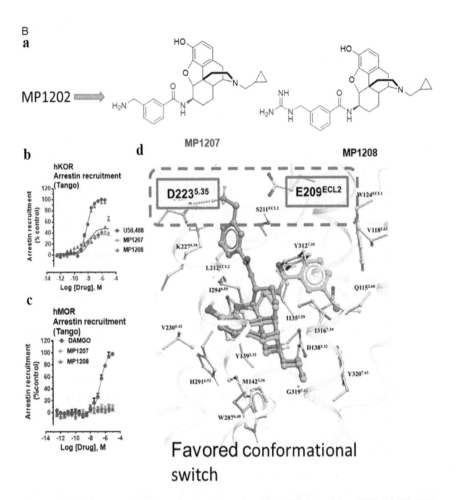

B
a

MP1202 ⟹

b

hKOR
Arrestin recruitment
(Tango)

- U50,488
- MP1207
- MP1208

c

hMOR
Arrestin recruitment
(Tango)

- DAMGO
- MP1207
- MP1208

d

MP1207

MP1208

D223^5.35

E209^ECL2

Favored conformational switch

Fig. 3—Cont'd recruitment assays compared to U50, 488 (purple) at KOR. (c) MP1207 (orange) and MP1208 (green) show no arrestin recruitment in Tango-arrestin recruitment assays compared to DAMGO (blue) at MOR. (d) Docking of MP1202 (green sticks) and MP1207 (yellow sticks) in wild type KOR shows MP1207 chair interact with D223 and E209 residues in TM5-ECL2 region by salt bridge formation, while MP1202 boat form does not accommodate this region. *Reproduced with permission from Uprety R, Che T, Zaidi SA, et al. Controlling opioid receptor functional selectivity by targeting distinct subpockets of the orthosteric site. Elife. 2021;10:e56519.*

in the TM2–TM3 region (Fig. 3Ae) and as a result, the compound retains arrestin recruitment. However, at MOR, the arm binds in the TM5–ECL2 region suggesting that this subpocket at the entrance of the orthosteric site may act as a hot spot to reduce arrestin signaling (Fig. 3Ad). To design KOR biased agonists, two acidic residues in the TM5-ECL2 region namely

D223$^{5.35}$ and E209^{ECL2} (Fig. 3Bd) were identified, and ligand design swapped the 3′-iodo group on the aryl ring with basic residues like 3′—NH$_2$ and 3′—NH—C=NH(NH$_2$) to force the 6β-amidophenyl arm to bind within the TM5-ECL2 region (Fig. 3Ba) and ensuring ring C takes a chair conformation at both receptors. Consistent with the design strategy, **MP1207** and **MP1208** showed vastly reduced arrestin recruitment at both MOR as well as KOR (Fig. 3Bbc). Analogs of **MP1207/1208** where the basic residue was substituted with a polar group like -3′OH behaved as balanced agonists while positioning the basic substituent (—NH$_2$) at the 4′ position on the phenyl ring again led to a loss of bias compared to **MP1207** as the salt bridge formation with the acidic residues was inhibited. The synthesis of MOR/KOR G-protein biased ligands targeting the TM5-ECL2 region using structure-based design is an example where chemists have taken advantage in rationally designing G-protein biased agonists, and this avenue could possibly be used at other targets as well. It must be noted that multiple hot spots at opioid receptors may exist beyond the TM5 ECL2 region which was exploited in the present case. It is also possible that this phenomenon of driving a reduction in arrestin efficacy is template-dependent.

5. Assay limitations in correlating *in vitro–in vivo* function

Receptor reserve refers to the phenomenon in which an agonist produces the maximal response by activating only a fraction of the available receptor population present in the system. For second–messenger systems that measure downstream G-protein signaling, signal amplification may yield a similar maximal response making it more difficult to differentiate between a full agonist and a partial agonist.[86,87]

The concept of using the right assay for assessing partial vs full agonists is important to distinguish between truly biased agonists and low efficacy G-protein biased agonists (see graphic presented in Fig. 4A–C). PZM21,[35] TRV130,[39] 7OH mitragynine[53,54,58] as well as SR17018[88] were previously described as G-protein biased agonists (described in this review under MOR biased agonists). However, when assessed in assays with minimum receptor amplification (assay was initially described by Stoeber et al.[89]and has recently been rigorously validated by Gillis et al.[30]), these compounds were found to be low efficacy agonists.[30,33] This calls for the design and development of compounds with either greater G-protein bias

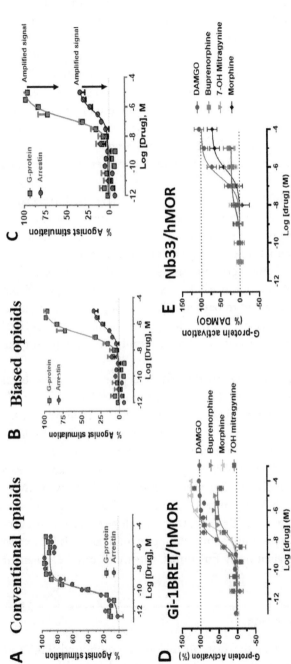

Fig. 4 Signaling differences in conventional, biased and partial opioids (A) Signaling behavior of conventional opioids. (B) Signaling behavior of biased opioids in an amplified assay. (C) Decrease in G-protein and arrestin efficacy of ligands with limited receptor reserve. (D) Buprenorphine and 7-OH Mitragynine are partial agonist with lower efficacy compared to DAMGO and Morphine in Gi-1 BRET assay. (E) Intrinsic efficacy of buprenorphine, morphine as well as 7OH mitragynine is lowered in the NB33 recruitment assays measured using BRET assays in hMOR. *Reproduced with permission from Bhowmik S, Galeta J, Havel V, et al. Site selective C–H functionalization of mitragyna alkaloids reveals a molecular switch for tuning opioid receptor signaling efficacy. Nat Commun. 2021;12(1):1–14; Chakraborty S, DiBerto JF, Faouzi A, et al. A novel mitragynine analog with low-efficacy Mu opioid receptor agonism displays antinociception with attenuated adverse effects. J Med Chem. 2021;64(18):13873–13892.*

and/or ones that are potency-biased instead of efficacy-biased and preferably with high efficacy in both G-protein and the arrestin pathway. Morphine, a well-characterized partial agonist shows an efficacy of 97–123%[30,32] normalized against DAMGO in an amplified assay, while in less amplified assays (e.g. in Nb33 recruitment assays), the efficacy is around 70% (Fig. 4D and E).[25,30,33] This scenario is seen with the previously defined biased agonists as well with PZM21 showing 84–86% efficacy[30] in an amplified assay compared to 38% efficacy in the same Nb33 assay and 7OH mitragynine, whose efficacy in BRET based assays and $[^{35}S]$-GTPγgS assays was 51%[33,54] and 77%[58] respectively, had an efficacy similar to buprenorphine in less amplified assays[33].

In addition to the Nb33 assay, other investigators have used irreversible antagonists[90] like β FNA, β-chlornaltrexone, and methcinnamox to evaluate the influence of this inactivation on agonist dose-response curves and again, arrive at an accurate estimation of true receptor intrinsic efficacy.[34]

More recently, low efficacy agonism was assessed in the VTA region[91] of the brain using electrophysiology as another ex-vivo measure of quantifying efficacy.[25] DAMGO was found to have a higher efficacy over morphine as expected from cell-based assays, while mitragynine agonists like SC13[25] were found to have lower efficacy over both morphine and the prototypic agonist DAMGO.

In addition, genetic strategies to decrease receptor levels have been looked at with the biased agonist TRV130 using the MOR +/− mice.[92] TRV130 showed no tolerance in wild-type mice when administered for 10 days but this reduced tolerance was not observed in the homozygous MOR +/− mice where a tolerance developed within four injections.

Utilizing the appropriate *in vitro* assay along with *ex vivo* and *in vivo* approaches is critical for discovering novel entities which show pathway selectivity and distinguishing them from partial agonists. It is hoped that leads identified through iterative SAR approaches will be evaluated in a battery of these efficacy assays before administering them in animals for *in vivo* efficacy analysis and evaluating if functional selectivity can be obtained *in vivo*.

6. Future directions

The FDA approval of the first MOR-biased compound TRV130 (oliceridine, Olinvyk) will allow for testing the biased opioid hypothesis in humans.[38] However, the hypothesis of biased agonism needs to evolve beyond one G-protein (Gi-1) and one arrestin subtype (β-arrestin2) to include other Gα-subtypes. Specifically, opioids are known to couple to G_{i1}, G_{i2}, G_{i3}, G_{oA}, G_{oB}, and G_z in addition to $G_{usducin}$ in the case of KOR.[62] Although preliminary, opioids tend to have specificity for the G_z-isoform.[25,62,93] Interestingly, MOR agonists and KOR agonists show a preference for the Gz subtype with partial agonists (buprenorphine at MOR and pentazocine at KOR) showing higher potency as well as efficacy at this subtype.[25,94]

While the exact role of Gz in opioid physiology is unknown, anticonception of morphine in Gz KO mice is moderately sensitive to morphine antinociception.[95] Also, an increased tolerance and lethality[96] was observed with morphine in the same KO mice. As more compounds are screened across this paradigm, it is hoped that compounds with selectivity at other subtypes will be synthesized, and individual roles of these isoforms will be evaluated. Overall, this strategy may represent a novel avenue to developing safer analgesics.

G-protein-biased agonists, as well as partial agonists, are now appearing in the literature with structures, such as cryo-EM structures of PZM21, FH210,[83] and mitragynine pseudoindoxyl[97] at MOR are now available. It is expected that these structures will aid in the atomic level understanding of receptor efficacy, potency, and pathway selectivity. In addition, these structures will allow for the rational design of the next generation of highly biased tool compounds to probe opioid functional selectivity in the near future.

Acknowledgment

S.M. is supported by funds from NIH grants DA046487, and start-up funds from St. Louis College of Pharmacy and Washington University.

References

1. Gurba KN, Chaudhry R, Haroutounian S. Central neuropathic pain syndromes: current and emerging pharmacological strategies. *CNS Drugs*. 2022;36(5):483–516.
2. Arfè A, Scotti L, Varas-Lorenzo C, et al. Non-steroidal anti-inflammatory drugs and risk of heart failure in four European countries: nested case-control study. *BMJ*. 2016;354: i4857.
3. Matsui H, Shimokawa O, Kaneko T, Nagano Y, Rai K, Hyodo I. The pathophysiology of non-steroidal anti-inflammatory drug (NSAID)-induced mucosal injuries in stomach and small intestine. *J Clin Biochem Nutr*. 2011;48(2):107–111.
4. Bindu S, Mazumder S, Bandyopadhyay U. Non-steroidal anti-inflammatory drugs (NSAIDs) and organ damage: a current perspective. *Biochem Pharmacol*. 2020;180.

5. Kimura M, Eisenach JC, Hayashida KI. Gabapentin loses efficacy over time after nerve injury in rats: role of glutamate transporter-1 in the locus coeruleus. *Pain*. 2016;157 (9):2024.

6. Ferguson JM. SSRI antidepressant medications: adverse effects and tolerability. *Prim Care Companion J Clin Psychiatry*. 2001;3(1):22.

7. Volkow ND, Blanco C. The changing opioid crisis: development, challenges and opportunities HHS public access. *Mol Psychiatry*. 2021;26(1):218–233.

8. Varga B, Streicher JM, Majumdar S. Strategies towards safer opioid analgesics—a review of old and upcoming targets. *Br J Pharmacol*. 2021.

9. Bruchas MR, Roth BL. New technologies for elucidating opioid receptor function. *Trends Pharmacol Sci*. 2016;37(4):279–289.

10. Hilger D, Masureel M, Kobilka BK. Structure and dynamics of GPCR signaling complexes. *Nat Struct Mol Biol*. 2018;25(1):4–12.

11. Liu X, Xu X, Hilger D, et al. Structural insights into the process of GPCR-G protein complex formation. *Cell*. 2019;177(5):1243–1251.e12.

12. Deupi X, Kobilka B. Activation of G protein-coupled receptors. *Adv Protein Chem*. 2007;74:137–166.

13. Al-Hasani R, Bruchas MR. Molecular mechanisms of opioid receptor-dependent signaling and behavior. *Anesthesiology*. 2011;115(6):1.

14. Mores KL, Cassell RJ, van Rijn RM. Arrestin recruitment and signaling by G protein-coupled receptor heteromers. *Neuropharmacology*. 2019;152:15–21.

15. Rasmussen SGF, Devree BT, Zou Y, et al. Crystal structure of the B2Adrenergic receptor-Gs protein complex. *Nature*. 2011;477(7366):549.

16. Zhou XE, He Y, de Waal PW, et al. Identification of phosphorylation codes for arrestin recruitment by G protein-coupled receptors. *Cell*. 2017;170(3):457–469.e13.

17. Raehal KM, Schmid CL, Groer CE, Bohn LM. Functional selectivity at the μ-opioid receptor: implications for understanding opioid analgesia and tolerance. *Pharmacol Rev*. 2011;63(4):1001–1019.

18. Faouzi A, Varga BR, Majumdar S. Biased opioid ligands. *Molecules*. 2020;4257.

19. Wootten D, Christopoulos A, Marti-Solano M, Babu MM, Sexton PM. Mechanisms of signalling and biased agonism in G protein-coupled receptors. *Nat Rev Mol Cell Biol*. 2018;19(10):638–653.

20. Faouzi A, Uprety R, Gomes I, et al. Synthesis and pharmacology of a novel μ-δopioid receptor heteromer-selective agonist based on the carfentanyl template. *J Med Chem*. 2020;63(22):13618–13637.

21. Cuitavi J, Hipólito L, Canals M. The life cycle of the Mu-opioid receptor. *Trends Biochem Sci*. 2020;27.

22. Kliewer A, Schmiedel F, Sianati S, et al. Phosphorylation-deficient G-protein-biased μ-opioid receptors improve analgesia and diminish tolerance but Worsen opioid side effects. *Nat Commun*. 2019;10(1).

23. White KL, Robinson JE, Zhu H, et al. The G protein-biased κ-opioid receptor agonist RB-64 is analgesic with a unique spectrum of activities in vivo. *J Pharmacol Exp Ther Am Soc*. 2015;352:98–109.

24. Kliewer A, Gillis A, Hill R, et al. Morphine-Induced respiratory depression is independent of β-arrestin2 signalling. *Br J Pharmacol*. 2020;177(13):2923–2931.

25. Chakraborty S, DiBerto JF, Faouzi A, et al. A novel mitragynine analog with low-efficacy Mu opioid receptor agonism displays antinociception with attenuated adverse effects. *J Med Chem*. 2021;64(18):13873–13892.

26. White JM, Irvine RJ. Mechanisms of fatal opioid overdose. *Addiction*. 1999;94 (7):961–972.

27. Bohn LM, Gainetdinov RR, Lin FT, Lefkowitz RJ, Caron MG. μ-Opioid receptor desensitization by β-arrestin-2 determines morphine tolerance but not dependence. *Nature*. 2000;408(6813):720–723.

28. Bohn LM, Lefkowitz RJ, Gainetdinov RR, Peppel K, Caron MG, Lin FT. Enhanced morphine analgesia in mice lacking β-arrestin 2. *Science*. 1999;286(5449):2495–2498.
29. Raehal KM, Walker JKL, Bohn LM. Morphine side effects in β-arrestin 2 knockout mice. *J Pharmacol Exp Ther*. 2005;314(3):1195–1201.
30. Gillis A, Gondin AB, Kliewer A, et al. Low intrinsic efficacy for G protein activation can explain the improved side effect profiles of new opioid agonists. *Sci Signal*. 2020;13(625):31.
31. Cuitavi J, Hipólito L, Canals M. The life cycle of the mu-opioid receptor. *Trends Biochem Sci*. 2021;46(4):315–328.
32. Uprety R, Che T, Zaidi SA, et al. Controlling opioid receptor functional selectivity by targeting distinct subpockets of the orthosteric site. *Elife*. 2021;10:e56519.
33. Bhowmik S, Galeta J, Havel V, et al. Site selective C–H functionalization of mitragyna alkaloids reveals a molecular switch for tuning opioid receptor signaling efficacy. *Nat Commun*. 2021;12(1):1–14.
34. Gillis A, Sreenivasan V, Christie MJ. Intrinsic efficacy of opioid ligands and its importance for apparent bias, operational analysis, and therapeutic window. *Mol Pharmacol Am Soc Pharmacol Exp Ther*. 2020;1:410–424.
35. Manglik A, Lin H, Aryal DK, et al. Structure-based discovery of opioid analgesics with reduced side effects. *Nature*. 2016;537(7619):185–190.
36. Kudla L, Bugno R, Skupio U, et al. Functional characterization of a novel opioid, PZM21, and its influence on behavioural responses to morphine. *Br J Pharmacol*. 2019. https://doi.org/10.1111/bph.14805.
37. Ding H, Kiguchi N, Perrey DA, et al. Antinociceptive, reinforcing, and pruritic effects of the G-protein signalling-biased Mu opioid receptor agonist PZM21 in non-human primates. *Br J Anaesth*. 2020;125(4):596–604.
38. Office of the Commissioner. *FDA approves new opioid for intravenous use in hospitals, other controlled clinical settings*; 2020. Available at: https://www.fda.gov/news-events/press-announcements/fda-approves-new-opioid-intravenous-use-hospitals-other-controlled-clinical-settings Accessed 7 September 2020.
39. DeWire SM, Yamashita DS, Rominger DH, et al. A G-protein-biased ligand at the μ-opioid receptor is potently analgesic with reduced gastrointestinal and respiratory dysfunction compared with morphines. *J Pharmacol Exp Ther*. 2013;344(3):708–717.
40. Yang Y, Wang Y, Zuo A, et al. Synthesis, biological, and structural explorations of a series of μ-opioid receptor (MOR) agonists with high G protein signaling bias. *Eur J Med Chem*. 2022;228.
41. Shulman M, Wai JM, Nunes EV. Buprenorphine treatment for opioid use disorder: an overview. *CNS Drugs*. 2019;33(6):567.
42. Grinnell SG, Ansonoff M, Marrone GF, et al. Mediation of buprenorphine analgesia by a combination of traditional and truncated mu opioid receptor splice variants. *Synapse*. 2016;70(10):395–407.
43. Khroyan TV, Polgar WE, Jiang F, Zaveri NT, Toll L. Nociceptin/orphanin FQ receptor activation attenuates antinociception induced by mixed nociceptin/orphanin FQ/μ-opioid receptor agonists. *J Pharmacol Exp Ther*. 2009;331(3):946–953.
44. Pergolizzi J, Aloisi AM, Dahan A, et al. Current knowledge of buprenorphine and its unique pharmacological profile. *Pain Pract*. 2010;10(5):428–450.
45. Dahan A. Opioid-induced respiratory effects: new data on buprenorphine. *Palliat Med*. 2006;20(suppl 1):3–8.
46. Canestrelli C, Marie N, Noble F. Rewarding or aversive effects of buprenorphine/naloxone combination (suboxone) depend on conditioning trial duration. *Int J Neuropsychopharmacol*. 2014;17(09):1367–1373.
47. Huang P, Kehner GB, Cowan A, Liu-Chen LY. Comparison of pharmacological activities of buprenorphine and norbuprenorphine: norbuprenorphine is a potent opioid agonist. *J Pharmacol Exp Ther*. 2001;297(2):688–695.

48. Alhaddad H, Cisternino S, Declèves X, et al. Respiratory toxicity of buprenorphine results from the blockage of P-glycoprotein-mediated efflux of norbuprenorphine at the blood-brain barrier in mice. *Crit Care Med*. 2012;40(12):3215–3223.

49. Brown SM, Holtzman M, Kim T, Kharasch ED. Buprenorphine metabolites, buprenorphine-3-glucuronide and norbuprenorphine-3-glucuronide, are biologically active. *Anesthesiology*. 2011;115(6):1251–1260.

50. Iribarne C, Picart D, Dréano Y, Bail JP, Berthou F. Involvement of cytochrome P450 3A4 in N-dealkylation of buprenorphine in human liver microsomes. *Life Sci*. 1997;60 (22):1953–1964.

51. Lutfy K, Eitan S, Bryant CD, et al. Buprenorphine-induced antinociception is mediated by Mu-opioid receptors and compromised by concomitant activation of opioid receptor-like receptors. *J Neurosci*. 2003;23(32):10331–10337.

52. Wilson LL, Chakraborty S, Eans SO, et al. Kratom alkaloids, natural and semi-synthetic, show less physical dependence and ameliorate opioid withdrawal. *Cell Mol Neurobiol*. 2021;41(5):1131–1143.

53. Chakraborty S, Majumdar S. Natural products for the treatment of pain: chemistry and pharmacology of Salvinorin A, mitragynine, and collybolide. *Biochemistry*. 2020;60 (18):1381–1400.

54. Kruegel AC, Gassaway MM, Kapoor A, et al. Synthetic and receptor signaling explorations of the mitragyna alkaloids: mitragynine as an atypical molecular framework for opioid receptor modulators. *J Am Chem Soc*. 2016;138(21):6754–6764.

55. Kruegel AC, Uprety R, Grinnell SG, et al. 7-Hydroxymitragynine is an active metabolite of mitragynine and a key mediator of its analgesic effects. *ACS Cent Sci*. 2019;5 (6):992–1001.

56. Gutridge AM, Chakraborty S, Varga BR, et al. Evaluation of kratom opioid derivatives as potential treatment option for alcohol use disorder. *Front Pharmacol*. 2021;12(11).

57. Gutridge AM, Robins MT, Cassell RJ, et al. G protein-biased kratom-alkaloids and synthetic carfentanil-amide opioids as potential treatments for alcohol use disorder. *Br J Pharmacol*. 2019;177(7):1497–1513.

58. Váradi A, Marrone GF, Palmer TC, et al. Mitragynine/corynantheidine pseudoindoxyls as opioid analgesics with Mu agonism and delta antagonism, which do not recruit β-arrestin-2. *J Med Chem*. 2016;59(18):8381–8397.

59. Zhou Y, Ramsey S, Provasi D, et al. Predicted mode of binding to and allosteric modulation of the μ-opioid receptor by Kratom's alkaloids with reported antinociception in vivo. *Biochemistry*. 2020;60(18):1420–1429.

60. Chakraborty S, Uprety R, Daibani AE, et al. Kratom alkaloids as probes for opioid receptor function: pharmacological characterization of minor indole and oxindole alkaloids from Kratom. *ACS Chem Neurosci*. 2021;12(14):2661–2678.

61. Chakraborty S, Uprety R, Slocum ST, et al. Oxidative metabolism as a modulator of Kratom's biological actions. *J Med Chem*. 2021;64(22):16553–16572.

62. Olsen RHJ, DiBerto JF, English JG, et al. TRUPATH, an open-source biosensor platform for interrogating the GPCR transducerome. *Nat Chem Biol*. 2020;16(8): 841–849.

63. Paton KF, Atigari DV, Kaska S, Prisinzano T, Kivell BM. Strategies for developing k opioid receptor agonists for the treatment of pain with fewer side effects. *J Pharmacol Exp Ther*. 2020;1:332–348.

64. Massaly N, Copits BA, Wilson-Poe AR, et al. Pain-induced negative affect is mediated via recruitment of the nucleus accumbens kappa opioid system. *Neuron*. 2019;102 (3):564–573.e6.

65. Bruchas MR, Schindler AG, Shankar H, et al. Selective P38α MAPK deletion in serotonergic neurons produces stress resilience in models of depression and addiction. *Neuron*. 2011;71(3):498–511.

66. Ehrich JM, Messinger DI, Knakal CR, et al. Kappa opioid receptor-induced aversion requires P38 MAPK activation in VTA dopamine neurons. *J Neurosci.* 2015;35 (37):12917.
67. Zhou L, Lovell KM, Frankowski KJ, et al. Development of functionally selective, small molecule agonists at kappa opioid receptors. *J Biol Chem.* 2013;288(51): 36703–36716.
68. Huskinson SL, Platt DM, Brasfield M, et al. Quantification of observable behaviors induced by typical and atypical kappa-opioid receptor agonists in male rhesus monkeys. *Psychopharmacology (Berl).* 2020;237(7):2075–2087.
69. Brust TF, Morgenweck J, Kim SA, et al. Biased agonists of the kappa opioid receptor suppress pain and itch without causing sedation or dysphoria. *Sci Signal.* 2016;9(456): ra117.
70. Schattauer SS, Kuhar JR, Song A, Chavkin C. Nalfurafine is a G-protein biased agonist having significantly greater bias at the human than rodent form of the kappa opioid receptor. *Cell Signal.* 2017;32:59.
71. Liu JJ, Chiu YT, DiMattio KM, et al. Phosphoproteomic approach for agonist-specific signaling in mouse brains: MTOR pathway is involved in κ opioid aversion. *Neuropsychopharmacology.* 2019;44(5):939–949.
72. Kaski SW, White AN, Gross JD, et al. Preclinical testing of nalfurafine as an opioid-sparing adjuvant that potentiates analgesia by the Mu opioid receptor-targeting agonist morphine. *J Pharmacol Exp Ther.* 2019;371(2):487–499.
73. Kumada H, Miyakawa H, Muramatsu T, et al. Efficacy of nalfurafine hydrochloride in patients with chronic liver disease with refractory pruritus: a randomized, double-blind trial. *Hepatol Res.* 2017;47(10):972–982.
74. Kumagai H, Ebata T, Takamori K, et al. Efficacy and safety of a novel κ-agonist for managing intractable pruritus in dialysis patients. *Am J Nephrol.* 2012;36(2):175–183.
75. Bohn LM, Aubé J. Seeking (and finding) biased ligands of the kappa opioid receptor. *ACS Med Chem Lett.* 2017;8(7):694–700.
76. White KL, Scopton AP, Rives M-L, et al. Identification of novel functionally selective K-opioid receptor scaffolds S. *Mol Pharmacol.* 2014;85:83–90.
77. Rives ML, Rossillo M, Liu-Chen LY, Javitch JA. 6′-Guanidinonaltrindole (6′-GNTI) is a G protein-biased κ-opioid receptor agonist that inhibits arrestin recruitment. *J Biol Chem.* 2012;287(32):27050–27054.
78. Ho JH, Stahl EL, Schmid CL, Scarry SM, Aubé J, Bohn LM. G Protein signaling-biased agonism at the k-opioid receptor is maintained in striatal neurons. *Sci Signal.* 2018;11:542.
79. Zangrandi L, Burtscher J, Mackay JP, Colmers WF, Schwarzer C. The G-protein biased partial κ opioid receptor agonist 6′-GNTI blocks hippocampal paroxysmal discharges without inducing aversion. *Br J Pharmacol.* 2016;173(11):1756–1767.
80. Fortin M, Degryse M, Petit F, Hunt PF. The dopamine D2 agonists RU 24213 and RU 24926 are also KAPPA-opioid receptor antagonists. *Neuropharmacology.* 1991;30 (4):409–412.
81. Spetea M, Eans SO, Ganno ML, et al. Selective κ receptor partial agonist HS666 produces potent antinociception without inducing aversion after i.c.v. administration in mice. *Br J Pharmacol.* 2017;174(15):2444–2456.
82. Erli F, Guerrieri E, Haddou TB, et al. Highly potent and selective new diphenethylamines interacting with the κ-opioid receptor: synthesis, pharmacology, and structure − activity relationships. *J Med Chem.* 2017;60:7579–7590.
83. Wang H, Hetzer F, Huang W, et al. Structure-based evolution of G protein-biased μ-opioid receptor agonists. *Angew Chem Int Ed Engl.* 2022.
84. Che T, Majumdar S, Zaidi SA, et al. Structure of the nanobody-stabilized active state of the kappa opioid receptor. *Cell.* 2018;172(1–2):55–67.e15.

85. Váradi A, Marrone GF, Eans SO, et al. Synthesis and characterization of a dual kappa-delta opioid receptor agonist analgesic blocking cocaine reward behavior. *ACS Chem Neurosci.* 2015;6(11):1813–1824.

86. Mcpherson J, Rivero G, Baptist M, et al. μ-Opioid receptors: correlation of agonist efficacy for signalling with ability to activate internalization. *Mol Pharmacol.* 2010; 78(4):756–766.

87. Kelly E. Efficacy and ligand bias at the μ-opioid receptor. *Br J Pharmacol.* 2013; 169(7):1430–1446.

88. Schmid CL, Kennedy NM, Ross NC, et al. Bias factor and therapeutic window correlate to predict safer opioid analgesics. *Cell.* 2017;171(5):1165–1175.e13.

89. Stoeber M, Jullié D, Li J, et al. Agonist-selective recruitment of engineered protein probes and of GRK2 by opioid receptors in living cells. *Elife.* 2020;9:e54208.

90. Vandeputte MM, Vasudevan L, Stove CP. In vitro functional assays as a tool to study new synthetic opioids at the μ-opioid receptor: potential pitfalls and progress. *Pharmacol Ther.* 2022;235:108161.

91. Margolis EB, Fields HL, Hjelmstad GO, Mitchell JM. δ-Opioid receptor expression in the ventral tegmental area protects against elevated alcohol consumption. *J Neurosci.* 2008;28(48):12672–12681.

92. Singleton S, Baptista-Hon DT, Edelsten E, McCaughey KS, Camplisson E, Hales TG. TRV130 partial agonism and capacity to induce anti-nociceptive tolerance revealed through reducing available M-opioid receptor number. *Br J Pharmacol.* 2021;bph.15409.

93. Masuho I, Ostrovskaya O, Kramer GM, Jones CD, Xie K, Martemyanov KA. Distinct profiles of functional discrimination among G proteins determine the actions of G protein-coupled receptors. *Sci Signal.* 2015;8(405).

94. Barnett ME, Knapp BI, Bidlack JM. Unique pharmacological properties of the kappa opioid receptor signaling through gaz as shown with bioluminescence resonance energy transfer. *Mol Pharmacol.* 2020;98(4):462–474.

95. Yang J, Wu J, Kowalska AM, et al. Loss of signaling through the G protein, Gz, results in abnormal platelet activation and altered responses to psychoactive drugs. *Proc Natl Acad Sci USA.* 2000;97(18):9984–9989.

96. Leck KJ, Bartlett SE, Smith MT, et al. Deletion of guanine nucleotide binding protein Az subunit in mice induces a gene dose dependent tolerance to morphine. *Neuropharmacology.* 2004;46(6):836–846.

97. Qu Q, Huang W, Aydin D, et al. Structural insights into distinct signaling profiles of the MOR activated by diverse agonists. *bioRxiv.* 2021. https://doi.org/10.1101/2021.12.07. 471645.

Index

Note: Page numbers followed by *"f"* indicate figures and *"t"* indicate tables.

Lightning Source UK Ltd.
Milton Keynes UK
UKHW022137060223
416587UK00001B/2